ENVIRONMENTAL AND WORKPLACE SAFETY

A Guide for University, Hospital, and School Managers

ENVIRONMENTAL AND WORKPLACE SAFETY

A Guide for University, Hospital, and School Managers

James T. O'Reilly
University of Cincinnati

Philip Hagan
Georgetown University

Peter de la Cruz and
the attorneys of
Keller & Heckman
Washington, D.C.

I(T)P® A division of International Thomson Publishing, Inc.
The ITP logo is a trademark under license

For more information, contact:

Van Nostrand Reinhold
115 Fifth Avenue
New York, NY 10003

Chapman & Hall GmbH
Pappelallee 3
69469 Weinheim
Germany

Chapman & Hall
2-6 Boundary Row
London
SE1 8HN
United Kingdom

International Thomson Publishing Asia
221 Henderson Road #05-10
Henderson Building
Singapore 0315

Thomas Nelson Australia
102 Dodds Street
South Melbourne, 3205
Victoria, Australia

International Thomson Publishing Japan
Hirakawacho Kyowa Building, 3F
2-2-1 Hirakawacho
Chiyoda-ku, 102 Tokyo
Japan

Nelson Canada
1120 Birchmount Road
Scarborough, Ontario
Canada M1K 5G4

International Thomson Editores
Campos Eliseos 385, Piso 7
Col. Polanco
11560 Mexico D.F. Mexico

1 2 3 4 5 6 7 8 9 10   BBR   01 00 99 98 97 96

Library of Congress Cataloging-in-Publication Data

O'Reilly, James T.
    Environmental and workplace safety : a guide for university,
hospital, and school managers / James T. O'Reilly, Philip Hagan,
Peter de la Cruz.
        p.   cm.
    Includes bibliographical references and index.
    ISBN 0-442-02123-2 (hardcover)
    1. Industrial safety—United States. 2. Industrial safety—Law
and legislation—United States. 3. Universities and colleges—
United States—Safety measures. 4. Hospitals—United States—
Safety measures. I. Hagan, Philip. II. De la Cruz, Peter.
III. Title.
T55.0683 1996
371.6′068′4—dc20   95-38757
                                                        CIP

*To Timothy James O'Reilly
with Love and Best Wishes
for His Growth and Success*

Contents

Preface

Banners on the field house rafters memorialize tournament champions. Bronze plaques in the hospital's main corridor honor the prizes won by outstanding researchers. A glass cabinet outside the president's office holds the NCAA Final Four trophy and a replica of the Pulitzer Prize awarded to an eminent professor. The largest of the institution's buildings are dedicated to honor wise professors' inventions, creations, and donations (sometimes the latter more than the former!). Educational and health-care institutions are great believers in awards and honors.

Somewhere in a small, cluttered office, perhaps in a far corner of the grounds or deep in a subbasement, the real winner devotes his or her time to the daily victories over risk that keep the institution thriving. The Safety Manager, Environmental Director, or Manager of Services, by whatever title described, merits a great deal more attention than the institution tends to give. The authors of this book want to recognize and applaud this professional for the many unsung contributions that keep the institution free of avoidable injuries, serious property damage, and environmental threats.

Our effort in this book is to make the hospital, university, school, or other institutions' manager aware of the full set of rules and laws that are imposed on the institution by government agencies relating to health, safety, and the environment. Awareness and evaluation of these commands is the best way by which to succeed in the safety and environmental role.

This text brings together the years of experience of several career professionals, including a team of environmental law experts and a team of safety managers from major institutions. Georgetown University Director of Safety Philip Hagan and veteran environmental scholar James O'Reilly developed the concept and enlisted the capable assistance of Keller & Heckman, one of the premier Washington law firms specializing in health, safety, and environmental law.

The attorneys of the eminent Washington law firm of Keller & Heckman, and especially partner Peter de la Cruz, deserve special praise for the research, writing, and creativity that have contributed so much to this book. Among the chapter contributors are David Sarvadi (workplace safety), Thomas C. Berger (water), Rosemarie A. Kelley (air), Martha E. Marrapese (waste management), Patrick W. Ratkowski (water, air, art, and theater hazards), and Carol M. Toth (hazardous-materials transportation). Their expert knowledge will make the reader well equipped for these issues.

Mr. Hagan warmly appreciates the help and support of his Georgetown University colleagues, including Jon Miller, whose recycling suggestions appear in Chapter 10, and the helpful ideas received from many peer university professional managers, including Ms. Janice Bradley of Rockefeller University and Mr. Lawrence Gibbs of Stanford University. Their input was invaluable.

Professor O'Reilly melded these many streams of ideas into the comprehensive text. In that endeavor, he thankfully acknowledges the excellent peer review insights of Pamela P. Gelbert of Randolph, New Jersey, and Barbara Keyes of the University of Cincinnati's Safety Department. Superior student researchers Kristin Morgan Tracy ('94) and Emily Briscoe ('95) helped make this book successful through outstanding legal research.

Although this book cannot replace the timely legal advice of the institution's legal counsel, we hope to offer ideas from which prompt and proper compliance planning can be achieved. The full answer to any specific question will come from the reader's own technical and regulatory experiences, as well as the consultation with and advice from legal counsel.

We hope that this book enhances your professional skills and stimulates further ideas from you for future evolution of this text in later years. We encourage readers with suggestions to write to Professor O'Reilly in care of Editorial Department, Van Nostrand Reinhold, 115 Fifth Avenue, New York NY 10003.

James T. O'Reilly
Cincinnati, Ohio

Chapter 1

Introduction

This is a book designed for problem solvers. By the nature of their career choice, institutional safety and environmental managers are solvers of complex problems within complex institutions. However, their worries are expanding as the reach of litigation, regulation, and liability is growing. The risk of personal exposure to liability and prosecution for environmental or workplace violations is a troubling part of the life of institutional facility and safety managers today. That risk arises out of the multiple interactions between government, the non-profit institution, and the individual manager.

Personal risks are a reason for the reader to pay attention to the lessons that this book teaches. The compliance problems exist, even if your institution has not yet officially confronted them. You are tasked with solving the problems of environmental and safety compliance and risk management, on a limited budget, and with signing the government forms that promise the institution's compliance with commands for workplace and environmental safety.

This text explains what the regulatory controls and litigation concerns will require of you, but it does not stop there. It is also an advocacy tool to help you show how a well-staffed and well-managed function will meet your institution's needs. By the term "advocacy," we mean the constant intra-institution pull and push over limited institutional budgets. If your role is counted as "overhead," your group may feel like an endangered species in the budget climate of the 1990s. Please incorporate the lessons from this book into your next dialogue about resources, staffing, and equipment.

Chapters 2 and 3 define the statutory and legal situation and the key terms and coverages that are relevant to the institution and its managers. A curious misunderstanding may exist at senior management levels, that nonprofits are excused from more burdensome compliance. If you have not already overcome that error, it may be holding back the funding and personnel that your institution provides for the safety and environmental function.

Chapter 4 shows how a major safety interaction—building-related risk issues—merits the greatest immediate attention, because it is the most frequent source of vulnerabilities for educational and health-care institutions. The regulatory schemes for air, water, waste, and so on, which are described in the chapters that follow, are thoroughly and accurately explained, so that a manager can become familiar enough to be sensitive to the regulatory commands.

Mandates of compliance with each of these schemes are challenges for the institution's manager. Personal liability is an increasing area of concern for those who certify air-emission permits or who sign waste reports. At the core of these chapters is the reader's recognition that, although the laws were passed with industry in mind, the scope now reaches virtually all institutions, down to the school and clinic levels.

Apart from these chapters on *regulation*, the text also contains important suggested guidance on *organization* of the institution's safety and environmental program. Leaders in this field are among the authors whose work contributed to this text. Experiences they have shared can be utilized as you assess what needs to be done to bring your functional group into twenty-first-century conditions of excellence. Feel free to create the structure that works best for you, rather than following any particular plan. But to the extent fiscal austerity in your institution is indirectly starving the safety protection of its employees, students, neighbors, or the environment, we urge that this text be used to make the case for better support and more consistent funding.

At the end of the text we cover some of the zoning and expansion problems that have been felt acutely by larger urban institutions. Neighbor opposition to traffic, noise, and truck loading is a part of the modern urban scene. In Chapter 15, as with others, the reader is encouraged to become proactive in the anticipation and avoidance of problems with expansion or new construction.

Sources beyond this text are available if you know where to reach for them. Many institutions have legal counsel on staff, but the specific regulatory commands may be beyond the routine mastery of university or hospital counsels, and these issues may be supported by advice from outside law firms. The manager of safety and environmental compliance should be an informed consumer of legal advice. Not all law firms have the expertise that these statutes require, so the manager who is in jeopardy should encourage the institution's counsel to seek the best expert advice available. Nationwide specialist firms like Keller & Heckman of Washington and others in a handful of cities have enjoyed exceptional growth from their aid to nonprofits who have been forced to deal with narrowly focused and aggressive federal enforcers.

Additional advice may come from one's professional society and its publications. The National Association of College and University Business Officers, the American Society of Safety Engineers, and other organizations have useful information to share. Industrial hygiene, environmental management, and regulatory-affairs organizations have periodicals with helpful articles, some of which are cited as references in this text.

If you need to interact with government organizations, the authors suggest that you "think local." There is likely to be an office of an air or water pollution agency close to your campus or home city. After studying this text and consulting with your legal counsel, call the agency for an appointment. Bring in a list of the questions you need to answer, and start a cordial dialogue if you can. These local people have local roots, and often have had connections to your institution. Let them know that you want to be in compliance and that you earnestly seek cooperation and guidance.

Why not fly immediately to Washington and seek the federal officials' specific directives? From the mountaintop, all trees look alike. From the central core of the bureaucracy, any institution would get the same set of forceful commands. It makes more sense to bring your inquiries to the regional or local office of the environmental or workplace safety agency and, if the answers are not readily available, ask for their help in seeking the interpretation or answer that you need. The note of feasibility and empathy with which they address the question locally is certain to exceed any response you would receive from Washington.

A final introductory note is in order, one that may be superfluous for many veterans of regulatory compliance. *You are what you document.* The records you keep and the correspondence you send determine the way you will be judged. Pay close personal attention to what you individually sign on behalf of your institution. If in doubt, confirm before signing. The ability of a federal prosecutor to use your documents against you is a painful reality for the defendant in an environmental crimes case. Jurors in a toxics-injury case, years after the alleged "exposure" began, will be reading the hasty E-mail message with which you warned a research lab manager about consequences you foresaw from sloppy actions in the lab or in its waste stream. Understand documentation and pay attention to what you sign, if you aspire to someday be a happy long-term retiree, not a dismayed defendant. Good luck!

Chapter 2

Is My Institution Covered by These Requirements?

2.1 THE SIGNIFICANCE OF AN INSTITUTION'S LEGAL STATUS

You know that environmental and workplace laws can be onerous, and you have read about their costs to factories and private workplaces, but you may have wondered: Do these laws apply to my school, hospital, or other institution? Should I be concerned?

In general, the answer is "yes," environmental laws apply to every "person," and the term "person" covers virtually all private- and public-sector entities. Under the law, a university is a "person"; a nonprofit or proprietary hospital is a "person"; a local school district is a "person." In some cases the statutes grant specific exceptions, such as a state law that differentiates between the state's own institutions and those of the regulated private-sector entities, or a Clean Air Act (CAA) exception for certain nonprofit entities (*Town of Brookline* v *Gorsuch* 1981). But for air, water, and waste laws generally, one should assume that they will apply to institutional entities, both nonprofits and proprietary institutions. To the extent that Congress adopts laws in order to regulate interstate commerce, the broad term "commerce" will reach into many sizes and shapes of educational and medical activities (*Marshall* v *Baptist Hospital* 1979; *NLRB* v *Catholic Bishop of Chicago* 1979).

There are exceptions in certain environmental laws that allow a school or hospital to avoid coverage. Some laws focused their environmental controls to impact on areas of highest priority, such as plants making explosives. One example is the community disclosure rules of the 1986 Emergency Planning and Community Right to Know Act (EPCRA). These chemical emission disclosure

rules are targeted at high-risk factories and refineries, those listed in certain manufacturing codes that are identified with manufacturing under the Census
Bureau's Standard Industrial Classification (SIC) codes. The factories within
these SIC codes must file toxic release inventory reports that list where releases
of chemicals have been made during the preceding calendar year. Hospitals and
schools are not covered. The inventory will not list any releases from an educational institution because coverage reaches to a limited set of "persons" whose
end product is manufactured goods, not healed patients or educated students. This
saves the institution both the costs and potential penalties of complying with the
EPCRA system. (In some states, state institutions have been included in a state
parallel system that reaches beyond EPCRA's coverages.)

Another example is the "facility" definition in solid-waste laws. The
Comprehensive Environmental Response, Compensation & Liability Act
(CERCLA, also called "Superfund") applies to a facility (Mays 1993). In some
cases, wastes from a hospital or academic institution are wastes from a "facility"
and a cleanup of waste can be required by the university (*U.S.* v *Alcan Aluminum*
1993; *Martin* v *Kansas Board of Regents* 1991).

In most situations, the institution's waste is unlikely to be closely regulated
because of the assumption that most wastes from a college or hospital would be
comparable to residential and office ("domestic") wastes rather than industrial
wastes. Ordinary solid waste could be regulated by the Environmental Protecton
Agency (EPA) but, except for incineration, the EPA rules have had only a modest effect on household garbage. The EPA has estimated that 4,000 metric tons of
hazardous wastes are generated by research and educational institutions each
year, scattered around the nation, with an EPA survey that found two large universities having 2,000 or more chemicals and mixtures for disposal each year
(EPA Guide 1990, 5). That number is likely to be too low an estimate, in light of
experience among the authors of this text. The local waste-planning districts
probably do not expect much risk from the college or hospital, though federal
experts have an expectation of a waste stream with a somewhat more likely hazardous nature.

Another categorization of regulated entities is made to determine
Occupational Safety and Health Act (OSHA) coverage. This category is
"employers" who are subject to OSHA regulation (University of Pittsburgh 1980;
Cochran v *International Harvester* 1975). State governments, their operating
entities, and political subdivisions are excluded from federal OSHA coverage, as
are employers of fewer than ten employees.

Virtually all hospitals, universities, and schools will be subjected to OSHA
inspections under either the federal OSHA or under state-law equivalent systems
(*Dept. of Labor* v *Dirt & Aggregate Inc.* 1992). A state hospital that is not covered by federal OSHA, for example, is probably covered by a state occupational-
safety scheme that protects state employees working in state-owned workplaces.
The federal OSHA standard that requires worker notification and training about

workplace chemical hazards has also been followed in many state programs affecting state employees. Only those state rules that conflict with or undermine the effect of federal OSHA rules are preempted by the federal system (*Gade* v *National Solid Waste Mgmt Assn.* 1992). Some 26 states have separate "state plans" that operate separately from the federal system, though they follow federal basic rules of compliance. Case-by-case decisions about exempt status can be expected. When the federal OSHA Review Commission considered whether the University of Pittsburgh was subject to federal OSHA or only to state laws, it ruled that federal OSHA applied where the university's special independent status did not give the university such public control, accountability, and functioning that would make it similar to the conditions of other state entities (University of Pittsburgh 1980).

2.2   PUBLIC INSTITUTIONS' STATUS

The Tenth Amendment to the Constitution restricts the federal government's ability to control the activities of the states, and the Eleventh Amendment restricts suits against the states in federal courts (*Pennsylvania* v *Union Gas Co.* 1989). However, state organizations (and, by legal derivation, county and municipal organizations) can be controlled by federal laws. This control applies where the activity of the state entity affects "commerce," through the same means that activities of other entities do, such as overtime pay for public workers (*Garcia* v *San Antonio MTA* 1985). For example, water-pollution and air-pollution laws regulate public-sector hospitals, colleges, and other entities with controls comparable to those that impact private sector buildings and factories.

The institution's tie to the public interest and civic responsibility does not excuse noncompliance, and might even increase its managers' vulnerability to criminal convictions or injunctions. Managers of city sewer plants and other publicly owned treatment works are faced with the same risks, penalties, and potential criminal charges that a private company discharging industrial wastewater would face. In the 1993 *Weitzenhoff* decision, for example, an appellate court upheld criminal convictions of city sewer district managers who diverted sewage into ocean disposal. The same conduct would have been criminally prosecuted in any of a hundred private-sector examples. There is little incentive to "give slack" to violators of environmental laws that cause environmental damage just because their workplace was not for profit.

Environmental managers of a publicly owned institution such as a college or hospital should expect that federal standards will apply. In the instances in which federal systems exclude state entities, each state is likely to have a state parallel statute with comparable, if not exactly similar, regulations. Federal OSHA regulations exclude states (University of Pittsburgh 1980) but the states have the option to include these entities under their own state laws; about half of the states administer both federal and state standards as what are called "state plan" states.

So, the potentially vulnerable manager asks: Is this entity owned or controlled by state government? The intricacies of state control can be daunting, but if there were a finding of exemption from federal regulations, premised on state government's exempted status, the classification effort might be helpful. Exclusion from federal labor controls does not exclude the institution or worker from appropriate state controls (*Amalgamated Assn.* v *Lockridge* 1971).

The need for federal-state coexistence is also matched by state and local government accommodation. In zoning cases, state laws might relieve a state-affiliated institution from local zoning controls (Olivieri 1975). Primacy of state power over local authority is a controversial topic, because it shields state colleges or state hospitals from local veto of expansion plans for a new or expanded state institution. The state institution might be caught by the terms of a state law requiring environmental impact statements, as discussed in section 15.04.

Other institutions want status as a state entity in order to claim sovereign immunity (Note Defective Buildings 1991, 351, 371). In tort cases, those states that require claims to be processed by the government agency first, and then litigated in a statutory venue such as a court of claims, manage to control some of the vulnerabilities of their state-owned institutions. If the state law allowing limited jurisdiction for negligence liability excludes state institutions from liability for assault and other intentional torts (*Delaney* v *University of Houston* 1992; *Townsend* v *Memorial Medical Center* 1975), then the creative plaintiff's counsel will try to avoid dismissal by differentiating the circumstances of the particular incident as "accidental," for example, the failure to take a certain pre-accident physical precaution such as the repair of a dormitory lock (*Delaney* v *University of Houston* 1992; *Cutler* v *Board of Regents* 1984).

2.3 NONPROFIT INSTITUTIONS STATUS

Status of the institution as a nonprofit is rarely a basis for a defense against an environmental prosecution, though it may be useful in defense of other litigation (*Marshall* v *Baptist Hospital* 1979). The "person" definitions that trigger coverage, described above, do not exclude nonprofits. Colleges are vulnerable to environmental enforcement cases, as are hospitals. Some exceptions to this norm exist (*Town of Brookline* v *Gorsuch* 1981).

Nonprofit status offers three benefits to avoid penalties, once a violation has been detected by enforcers. *First*, the defendant's status diminishes the desire of prosecutors to punish violations that would have been front-page news if the violator were a chemical plant or refinery. Demanding a $100,000 penalty from a charity hospital for migrant workers that mishandles medical waste is not an attractive news media event for the zealous environmental prosecutor.

This mitigation factor can be financially very beneficial. In the *Florida Keys* case, the college argued that environmental adverse effects of its wetlands filling should be offset by the college's beneficial use of the wetlands for a public education purpose, and the court imposed a lessened level of mitigation against the defendant (*U.S.* v *Board of Trustees of Florida Keys* 1981).

To be clear, nonprofits benefit from this mitigation without having a revival of the obsolete "charitable immunity" doctrine (*Mink* v *University of Chicago* 1978; Note Defective Buildings 1976, 351, 376). The nonprofit nature of the college or hospital makes it less likely to be hit for punitive damages, but this is an element that may vary in particular cases when considered in light of active involvement with the injurious conduct, negligence, size of endowment, and income, availability of insurance, and so on (*Mink* v *University of Chicago* 1978).

A *second* advantage of nonprofit status is that the environmental enforcement agencies use computer models to determine the economic benefit that private-sector companies derived from the violation. Equalizing the consequences of compliance with those of not complying makes sense (Campbell-Mohn 1993, 178). This "BEN Modeling" is a measure of profits, productivity, and rates of return on investment that calculates an assumed amount of benefit upon which a penalty will be calculated. So, a nonprofit that has violated an environmental law is likely to face a lessened penalty for the same quantity of waste that would have generated a higher computed penalty under the BEN process where the defendant was a manufacturer.

A *third* benefit of nonprofit status is the probable reluctance of courts to award punitive damages against nonprofits (*Sterner* v *Wesley College* 1990). This inclination will vary with the circumstances of each case.

2.4   OTHER TYPES OF LIABILITY EXPOSURE

Beyond status as a legal "person," and beyond exceptions and exclusions in regulatory laws, the institution also can be bound to pay damages for accidents or environmental problems under several different legal approaches.

The employer-employee relationship may make the institution liable under two subcategories of state workers' compensation laws. The workers' compensation system is "exclusive" and, once employee status is established, generally bars the individual from suing his or her employer for injuries that occurred on the job or during the course of work activity (Larsen 1994; Note Workers 1984). In some states, violation of a specific safety rule that applies to this type of work may result in a doubling of the workers compensation benefit payments, as a means of deterring employers from violations of applicable safety requirements. In Ohio, for example, the injured worker can collect an extra cash benefit if the workers' compensation board finds that the employer's supervisor had ordered the worker to perform an activity without mandatory safety equipment, in violation of workplace safety rules. The violation must have been of a specific safety rule that was applicable at the time of the injury. A natural question is the liability of the institution when an employee is the source of the violation, for example, failure to wear personal protective equipment (PPE). In rare cases OSHA has excused the violation when the employer has documented that it has a system in place that requires PPE to be worn, and that the employee was subject to discipline for breach of the safety rules. The better the documentation, the more likely that this defense claim of employee breach of a safety rule may be accepted.

In many states, "intentional tort" laws or case precedents permit the individual to sue the employer for intentionally ordering the employee to perform an action where the employer had knowledge of a very dangerous condition, directed the employee to work under that condition, and either intended injury would occur or knew that injury was substantially likely to occur, beyond the standards of negligence or recklessness (Larsen 1994).

The duties of the institution as a landlord affect safety-related liabilities (Flaherty 1988); some states impose liability for dormitory assaults (*Nero* v *University of Kansas* 1993) while other states do not (*Savannah College* v *Roe* 1991). In the college's role as the landlord that rents dormitory rooms to students or the hospital's role of providing apartment housing for nurses, the responsible managers should check out and remedy unsafe conditions to lessen the institution's exposure to suits, but the institution is not acting in the place of parents. Colleges can be held liable for assaults (*Nieswand* v *Cornell University* 1988) or accidents (*Shetina* v *Ohio University* 1983) under state court precedents, applying the "common law" of responsibilities that has evolved over decades of landlord-tenant disputes. A tenant such as a resident student, or a "business invitee" (*Dickson* v *Yale University* 1954) such as a commuting student, may be able to win a negligence case if a fire, slip and fall, or other injury event occurs. Student families pose an additional route of liability; children and spouses injured on campus may be plaintiffs and might recover, in the same way that paying tenants could recover, damages against a negligent landlord. Responsibility of the college for actions of third parties such as student pranks (*Rabel* v *Illinois Wesleyan University* 1987) or a fraternity hazing group poses difficult questions of attributing liability (*Furek* v *University of Delaware* 1991).

The institution may have opened itself to liability for accidents under a theory of "warranty" defined in Uniform Commercial Code section 2-313, if its agreements or published documents make promises relating to safety or protection from external harms (*Nieswand* v *Cornell University* 1988; *Mullins* v *Pine Manor College* 1983). The typical example is a criminal assault that occurs at the institution, in which the victim is injured. Lawyers for the injured person look first to state laws that create specific rights or duties. These would be examined first to see if the college or hospital has special state-law exclusions from liability. The creative plaintiff's lawyer looks then for promises of safety or assurances of security that may have been written into housing brochures, rental contracts, or other communications between the institution and the injured person (*Nieswand* v *Cornell University* 1988; *Mullins* v *Pine Manor College* 1983).

   The goal of the challenger is to sue for breach of the promise that the person would be safe in the apartment or dormitory room. The "breach of warranty" claim requires the challenger to show that the promise of safety was issued by the institution; that the person relied on that promise when paying rent, room and board, and so on; and that the injury was a result of breach of that promise of safety. An unusual lawsuit in 1994 accused Albright College of breach of contract (premised on resident student selection materials) when the quiet, studious plaintiff was mismatched with a roommate who favored aggressive social adventures and the plaintiff dropped out after two months and sued for damages (Oil and Water 1994).

   Occasionally, special situations result in even further expansion of the coverage of special legislation and the institution becomes a "generator" or an "agency." Status as a "generator" under the federal Resource Conservation and Recovery Act (RCRA) or state medical waste laws, discussed in Chapter 7 of this text, carries special regulatory obligations to those who deal with waste from the facility (Menicucci and Coon 1994). When a university acts as a federal contractor, it is deemed to be part of an "agency" for purposes of the National Environmental Policy Act (NEPA) requirement to prepare an environmental impact statement. This means the university can be enjoined from operating a federally funded program without undergoing the delays and costs of performing an environmental impact statement (*Foundation* v *Heckler* 1985).

REFERENCES

Amalgamated Assn. of Street etc. Workers v *Lockridge*, 403 U.S. 274 (1971).
Campbell-Mohn, Cecilia. 1993. Objectives and Tools. In *Sustainable Environmental Law,* ed. Cecilia Campbell-Mohn, Barry Breen, and J. William Futrell. St. Paul, MN; West Publishing.
Cochran v *International Harvester,* 408 F. Supp. 598 (W.D. Ky., 1975).
Cutler v *Board of Regents,* 459 So.2d 413 (Fla. App. 1984).
Delaney v *University of Houston,* 835 S.W.2d 56 (Tex. 1992).
Department of Labor & Industries v *Dirt & Aggregate Inc.,* 120 Wash. 2d 49, 837 P.2d 1018 (1992).

Dickson v *Yale University,* 195 A.2d 463 (Conn. 1954).

Flaherty, Michael. 1988. Tort Liability of College or University for Injury Suffered by a Student. *American Law Reports 4th,* 62:81.

Foundation on Economic Trends v *Heckler,* 756 F.2d 143 (D.C. Cir., 1985).

Furek v *University of Delaware,* 594 A.2d 506 (Del. 1991).

Gade v *National Solid Waste Management Association,* 112 S.Ct. 2374 (1992).

Garcia v *San Antonio Metropolitan Transit Authority,* 469 U.S. 528 (1985).

Larsen, Arthur. 1994 Supp. *Law of Workers Compensation.* New York: Matthew Bender 2A:65.00.

Marshall v *Baptist Hospital Inc.,* 473 F. Supp. 465 (M.D. Tenn. 1979), revsd, 668 F.2d 235 (6th Cir., 1980).

Martin v *Kansas Board of Regents,* 1991 Westlaw 33602 (D. Kans. 1991).

Mays, Richard. 1993. *CERCLA Litigation, Enforcement & Compliance.* Colorado Springs: McGraw-Hill/Shepards.

Menicucci, Margaret, and Cheryl Coon. 1994. Environmental Regulation of Health Care Facilities: A Prescription for Compliance. *Southern Methodist University Law Review.* 47:537.

Mink v *University of Chicago,* 460 F.Supp. 713 (N.D. Ill. 1978)

Mullins v *Pine Manor College,* 449 N.E.2d 331 (Mass. 1983).

Nero v *University of Kansas,* 861 P.2d 768 (Kan. 1993).

Nieswand v *Cornell University,* 692 F.Supp. 1464 (N.D.N.Y. 1988).

NLRB v *Catholic Bishop of Chicago,* 440 U.S. 490 (1979).

Note (Student Work). 1991. Defective Buildings and Grounds—A Dangerous Condition for Colleges and Universities. *Journal of College and University Law* 17:351, 376.

Note (Student Work) 1984. Workers' Compensation and College Athletics: Should Universities Be Responsible for Athletes Who Incur Serious Injuries? *Journal of College and University Law* 10:197.

Oil and Water. 1994. *Insight Magazine* 10:34 (July 11, 1994).

Olivieri, Daniel. 1975. Zoning Regulations as Applied to Colleges, Universities or Similar Institutions for Higher Education. *American Law Reports* 3d 64:1138.

O'Reilly, James, Kyle Kane, and Mark Norman. 1993. *RCRA and Superfund Practice Guide.* 2d ed. Colorado Springs: McGraw-Hill/Shepards.

Pennsylvania v *Union Gas Co.,* 491 U.S. 1 (1989).

Rabel v *Illinois Wesleyan University,* 195 A.2d 463 (Conn. 1954).

Savannah College of Art & Design v *Roe,* 409 S.E.2d 848 (Ga. 1991).

Shetina v *Ohio University,* 459 N.E.2d 587 (Ohio App. 1983).

Sterner v *Wesley College,* 747 F. Supp. 263 (D. Del., 1990).

Town of Brookline v *Gorsuch and Harvard College,* 667 F.2d 215 (1st Cir., 1981).

Townsend v *Memorial Medical Center,* 529 S.W.2d 264 (Tex. Civ. App. 1975).

U.S. v *Alcan Aluminum and Cornell University,* 950 F.2d 711 (2d Cir., 1993).

U.S. v *Board of Trustees of Florida Keys Community College,* 531 F. Supp. 267, 272 (S.D. Fla. 1981).

U.S. v *Weitzenhoff,* 35 F.3d 1275 (9th Cir., 1993).

USC is 29 United States Code §651(6).

U.S. Environmental Protection Agency. 1990. *Guides to Pollution Prevention: Research and Educational Institutions* (EPA/625/7-90/010). Washington DC.

Uniform Commercial Code §2-313.

University of Pittsburgh. 1980. *Occupational Safety & Health Decisions (CCH)* para. 24,240 (Rev. Com. 1980).

Chapter 3

Which Persons Are Covered by Which Requirements?

3.1 DEFINING YOUR INSTITUTION'S "EMPLOYEES"

The simple term "employee" is a source of confusion in federal labor laws (*NLRB v Hearst* 1944), but it packs a lot of legal significance for the manager who is required to deal with workplace and environmental safety concerns (Commerce Clearing House Inc. 1995). Universities may have the most difficult time categorizing persons as "employees" because their intra-institutional mix of students, graduate assistants, work-study helpers, contractors, and part-time workers (*NLRB v Yeshiva University* 1980) exceeds in complexity even the physician and group-practice uncertainties with which hospital administrators must cope (Flaherty 1988). This section provides an explanation of the term "employee" that may clear up some of the uncertainty that the word engenders in today's complex labor and environmental rulings.

State laws and federal wage and labor statutes define which workers are "employees" for payroll purposes, but the safety-related status decision may require a slightly different evaluation. Workers' compensation rules of your state are the first place to check a person's disputed status (Larsen 1994). The multiphase set of criteria for "employee" status in your state's workers compensation decisions are likely to include the worker's supervision, power to terminate the worker, payment authority, discipline powers, and so on. For safety issues, these workers' compensation decisions are the primary reference point, though in some cases an interpretation based on federal decisions of the U.S. Department of Labor's Occupational Safety and Health Administration (*Secretary of Labor v Magor Plumbing* 1993) and its Wage and Hour Administration about "employee" status will be needed to determine the unusual cases.

The state administrative agency responsible for labor enforcement probably has a set of principles in a guideline that it will provide on request. The state agency that rules on disputed claims for unemployment insurance will have their own set, adopted under the federal unemployment compensation system. If the matter arises after an injury, then state workers' compensation board decisions and federal OSHA interpretations of "employee" status are likely to be controlling (USC, §652(6)).

Federal OSHA legislation excludes from federal OSHA's jurisdiction the employees of state and local government institutions, but a majority of these are covered by the equivalent standards and programs in state law. About half of the states have chosen to adopt federal OSHA standards as part of a "state plan" of enforcement, and in these states, the coverage of OSHA requirements is universal. The gap of protection may exist for workers of a state or federal entity in a state that is subject to federal control of private-sector workplaces, but that imposes no extra occupational safety controls upon public-sector workers.

Sometimes the "employee" definition is clouded by religious institutions' peculiarities. Though a Catholic diocesan hospital follows policies set by the Vatican, the "employer" is not the Pope, but the local diocesan bishop (*NLRB* v *Catholic Bishop of Chicago* 1979), and the individual employees are not employees of the Pope but of the diocese, or more frequently, of the hospital's non-profit corporation. Religious affiliation of a school, university, or hospital is not a basis for it to claim exclusion from safety and environmental regulation (Livingstone College 1987). While OSHA excludes churches in their religious capacity from coverage, and does not protect clergy, choir, organist, and so on, all the conventional worldly activities that are sponsored by a church will be subject to OSHA regulation.

The OSHA test for "employee" status, used by the independent Occupational Safety and Health Review Commission, determines whether a penalty will be imposed for unsafe working conditions, inadequate safety documentation, or other workplace violations. The OSHA "employee" standard looks at several factors: who controls the activities of the worker; who has the actual power to control the employee; and who has the power to fire the employee or modify employment conditions. Two lesser factors to be considered also include who the worker considers to be his/her employer, and who pays the wages. Private contractual arrangements are helpful but are not controlling on the OSHA adjudications. The OSHA protection of "employees" includes managers, and the president of the company in one case.

An additional source of guidance may be National Labor Relations Board (NLRB) rulings on voter eligibility in labor representation elections at which only "employees" may vote. One who wants to vote for a union representative may be opposed by the employer, who argues for independent contractor status (*Cochran* v *International Harvester* 1975) or for some other status as a way to exclude that ballot from the election. In the context of a health or safety problem, however, courts would look first at the safety-related agencies' interpretations of "employee" status, before ruling on coverage.

3.2 STATUS OF STUDENTS AND INTERNS

Universities and schools have students, teaching assistants, part-time coaches, and many other categories of persons who are exposed to potential risks. Many hospitals have interns as well as medical and nursing students who may be working within the institution. Do environmental protection and workplace safety laws apply to these persons as well as to "employees"?

Employees benefit from the workplace safety law requirement that there is a general duty of the employer to provide a safe workplace. (*Guy* v *Northwest Bible College* 1964). NLRB decisions provide a basis for concluding that paid medical interns and residents have "employee" status. Graduate teaching assistants are classified in labor law as "students" and not as "employees" (Leland Stanford University 1974).

A college's resident students are not classed as employees but as "business invitees" (Restatement of Torts 1965; *Leonardi v Bradley University* 1993; *Isaacson Husson College* 1972) or tenants (Nero 1993). Visitors who are not students or who do not have business reasons for their presence at the institution are "licensees" on campus (*Doelker* v *Ohio State University* 1990; *Siver* v *Atlantic Union College* 1958; Restatement of Torts 1965), so the law applies only a limited duty for the institution not to act wantonly or willfully with respect to risks of injury (*Doelker* v *Ohio State University* 1990). Trends of case decisions do not hold universities liable for injuries that occur when the injured plaintiff was intoxicated on campus.

Proximate cause issues become complex when unusual conduct of the student (*Weissman* v *State* 1987) or visitor contributes to the harm. Courts are usually skeptical of the defense by an institution that the injured person was solely liable for acting in a fully independent negligent manner. The person's injury might be attributed to the institution in the foreseeable harm case (*McIlrath* v *College of St. Catherine* 1987; *Pizzola* v *State* 1987). Colleges have no special obligation to protect students from the acts of other students, such as in the fraternity hazing setting, but some courts have applied specific liabilities to some colleges (*Peterson* v *San Francisco Community College* 1984). Facts of each case will be weighed, but the likely bias of the jury or administrative benefits decision maker will be to grant benefits (or damage awards) against the wealthy institution in favor of the poor student or bypasser (Note Liability 1989).

3.3 STATUS OF MANAGERS, FACULTY,
AND RESPONSIBLE OFFICIALS

Full-time faculty of a university are managers, not employees of the university, for purposes of federal labor law (*NLRB* v *Yeshiva University* 1980). For purposes of OSHA and environmental laws, the faculty, staff, and management of a university are "employees" and their actions that may violate the law will ordinarily be attributed to the institution.

In the hospital setting, managers or officials will routinely be treated as employees under the OSHA and environmental laws. The intricacies of contracts between a hospital and physical groups such as a contractual operator of an emergency room have not yet reached the appellate courts in reported cases under safety or environmental laws. But it is very unlikely that the institution and any individual who released a pollutant will be jointly charged, unless a prosecutor does so to force an individual to exculpate herself by blaming the institution.

When and if one of these liability issues arises, for example, if unmarked medical waste bags are dumped by physicians in a hospital's domestic waste dumpster and found by an environmental agency inspector, the question of the relative authority and control of the institution over the individual will be raised. The practice of not vigorously pursuing an individual factory employee in a joint institutional/individual complaint of environmental violations may have been the past practice of agencies, but where the individual is a sophisticated and well-paid physician, dentist, or professor, there is room for the agencies to pursue active cases against both institution and individual.

Tort law has a separate set of principles for the attribution of supervisors' actions to the institution. These conventional civil-damage principles apply, for example, when a supervisor uses due care in overseeing an employee whose misconduct injures a third party (*Malone* v *Miami University* 1993).

3.4   STATUS OF INDEPENDENT CONTRACTORS

The law of independent contractor liability seems to be one part labor law, one part tax law, and one part contract law (*Cochran* v *International Harvester* 1975). OSHA tends to study the economic realities of the particular work site, and penalize the institution that was in control of the workers who were exposed to risk (*Secretary of Labor* v *Magor Plumbing* 1993; Commerce Clearing House Inc. 1995). For example, during an investigation of a trench collapse or injuries to a hand in a drill press, the OSHA inspector will inquire about how the worker has been directed and supervised in performance of the worker's tasks.

If "co-employer" liability is found, then the payment of a penalty by both the institution and the contracting company may be required, since each had a responsibility to assure a safe workplace for the injured person (Commerce Clearing House Inc. 1995).

3.5   STATUS OF VISITORS AND BUSINESS "INVITEES"

Apart from environmental exposures and workplace risks, there are the conventional civil-damage cases that assert that harmful conduct or negligence resulted in physical injury to non-"employees" who entered into the university or hospital grounds. Persons who enter the university, school, or hospital may be injured accidentally and may sue for damages. "Premises liability" is the term used for this category of lawsuits. The legal standard for the trial judge's instructions to a civil jury, after a trial, is more favorable to the plaintiff if the reason the plaintiff was present at the institution had something to do with its mission or its business functions. Higher standards of legal duty of care are applied to those at the institution for business purposes, known as "business invitees" (Restatement of Torts 1965), than the legal duty of care owed to those who are casually present or present for their own purposes, known as "licensees."

Though the injured plaintiff bears a burden to show her status and thus the higher duty of care of the institution toward her, "invitee" is the legal category into which students, sales representatives, and persons generally associated with the business of the school will tend to be classified in a civil-damages case (*Curtis* v *State* 1986; *Mortiboys* v *St. Michael's College* 1973).

The determination of "invitee" or "licensee" status is a small fraction of the total case for civil damages, since it merely raises or lowers the legal standard that the court will use when instructing the jurors about the degree of care that the negligent university should have followed. In practice, most of these cases are settled and the courts handle these liabilities as they would for any nonacademic or non-health-care institution or shop owner (Note Defective Buildings 1991). The normal slip-and-fall case, for example, would presumably be covered by conventional insurance coverage of the university for premises liability (Higgins and Zulkey 1990).

3.6   LIABILITY OF INSTITUTION
AND INDIVIDUAL MANAGERS

Personal liability for environmental harms may take the form of civil administrative penalties, injunction orders, or even criminal penalties that name individual managers (*U.S.* v *Johnson & Towers* 1985). Because environmental laws allow enforcement agencies to apply penalties to certain individual managers as well as to institutions, it is important to understand how you and your colleagues might be held individually responsible for spills, releases, and other chemical incidents.

The first category of individual liability is fraud and concealment. The system of regulation runs on trust and verification. Individuals, especially managers, are penalized when they cheat the regulatory agency by filing false documents. Tough sentences under the federal sentencing guidelines and their state equivalents will be used to punish fraud (U.S. Sentencing Commission 1993).

The second category is cases of environmental offenses where the prosecution can show a manager's personal knowledge but only reckless action or inaction in the face of knowledge. Here the showing of a greater degree of education and experience will weigh against the individual. Prosecutors might reach to an environmental engineering department chair who knows that metalworking solvents require Resource Conservation and Recovery Act (RCRA) manifests, but who at semester's end goes on an extended vacation and leaves four drums outside the back door of the chemistry building without characterizing and documenting the waste. Situations vary considerably.

The third category is negligent actions so egregious that criminal negligence is charged by prosecutors. States can criminally charge a reckless employer even if the same action would have also violated workplace safety rules enforced by federal officials (*People* v *Chicago Magnet Wire* 1990). A criminal conviction of a small-business manager, who dumped hazardous waste into a normal solid-waste dumpster, occurred after local children playing in the area of the dumpster climbed into it and were killed by the vapors in the confined space (*U.S.* v *Johnson & Towers* 1985). This tragic scenario could occur where buildings that house, for example, chemical laboratories and early-childhood day-care training centers, are close together on a campus.

In general, penalties are assessed against the institution rather than the merely negligent manager. Janitors are seldom pursued or indicted. The case may arise where public abhorrence of the environmental offense leads to individual as well as group charges. Those college or hospital managers who have the professional training and expertise to have acted correctly are very likely to be named as co-defendants by an aggressive agency enforcement manager.

REFERENCES

Cochran v *International Harvester,* 408 F. Supp. 598 (W.D. Ky., 1975).

Commerce Clearing House Inc. 1995. *Employment Safety and Health Guide* Para. 509.

Curtis v *State,* 29 Ohio App.3d 297, 504 N.E.2d 1222 (1986).

Doelker v *Ohio State University,* 61 Ohio Misc. 2d 69, 573 N.E. 2d 809 (Ohio Ct. Claims, 1990).

Feld, Dan E. 1976. Who Is an "Employee" for Purposes of the Occupational Safety & Health Act? *American Law Reports Fed.* 27:943.

Flaherty, Michael. 1988. Tort Liability of College or University for Injury Suffered by Student as a Result of Own or Fellow Student Intoxication. *American Law Reports 4th* 62:81.

Gallagher, Richard. 1982. Hospital House Staff Physicians as "Employees." *American Law Reports Fed.,* 57:608.

Guy v *Northwest Bible College,* 64 Wash. 2d 116, 390 P.2d 708 (1964).

Higgins, Byron, and Edward Zulkey. 1990. Liability Insurance Coverage. *Journal of College and University Law* 17:123.

Isaacson v *Husson College,* 297 A.2d 98 (Maine 1972).

Larsen, Arthur. 1994 Supp. *The Law of Workers' Compensation.* New York: Matthew Bender. 1C: Chapter VIII; 2A:65.00.

The Leland Stanford University, 214 NLRB No. 621 (1974).

Leonardi v *Bradley University,* 625 N.E.2d 431 (Ill. App. 1993).

Livingstone College, 286 NLRB No. 124, 126 LRRM 1332 (1987).

Malone v *Miami University,* 89 Ohio App.3d 527, 625 N.E.2d 640 (1993).

McIlrath v *College of St. Catherine,* 399 N.W.2d 173 (Minn. App. 1987).

Mortiboys v *St. Michael's College,* 478 F.2d 196 (2d Cir., 1973).

Nero v *University of Kansas,* 861 P.2d 768 (Kan. 1993).

NLRB v *Catholic Bishop of Chicago,* 440 U.S. 490 (1979).

NLRB v *Hearst,* 322 U.S. 111 (1944).

NLRB v *Yeshiva University,* 444 U.S. 672 (1980).

Note (Student Work). 1991. Defective Buildings and Grounds—A Dangerous Condition for Colleges and Universities. *Journal of College and University Law* 17:351, 376.

Note (Student Work). 1989. Hazing and the "Rush" to Reform. *Journal of College and University Law* 16:93.

Note (Student Work). 1989. The Liability and Responsibility of Institutions of Higher Education for the On-Campus Victimization of Students. *Journal of College and University Law* 16:119.

People v *Chicago Magnet Wire Corp.,* 126 Ill.2d 356, 534 N.E.2d 962 (1989), cert. den. 110 S.Ct. 52 (1990).

Peterson v *San Francisco Community College District,* 36 Cal.3d 799, 685 P.2d 1193, 205 Cal. Rptr. 842 (1984).

Pizzola v *State,* 130 A.D.2d 796, 515 N.Y.S.2d 129 (1987).

Restatement of Torts 2d §342 (1965).

Secretary of Labor v *Magor Plumbing & Heating Co.,* 16 BNA Occup. Safety & Health Decisions 1223 (1993).

Siver v *Atlantic Union College,* 338 Mass. 212, 154 N.E.2d 360 (1958).

U.S. v *Johnson & Towers Inc.,* 741 F.2d 662 (3d Cir., 1984), cert. den. 469 U.S. 1208 (1985).

U.S. Sentencing Commission. 1993. *Federal Sentencing Guidelines,* 18 United States Code Appendix 4 at 2F1.

Weissman v *State,* 94 A.D.2d 796 (N.Y. 1987).

Chapter 4

Workplace and Building Safety

4.1 ASBESTOS CONTROL: INTRODUCTION

The building manager must learn about the control of asbestos and must "practice what we preach" in handling this costly and risk-related substance. Compliance with Occupational Safety and Health Administration (OSHA) asbestos standards is absolutely essential to a well-managed program. The basic message is:

> *In buildings constructed before 1981, federal rules presume that exposure to airborne asbestos fibers must be controlled, whenever workers contact, disturb or remove (1) thermal-system insulation, (2) tile flooring (asphalt, vinyl or resilient), or (3) surfacing materials that were sprayed or troweled on, such as plaster walls, soundproofed surfaces, or fireproofed surfaces.*

The model for asbestos management is taken from the 1986 Asbestos Hazard Emergency Response Act (AHERA), under which EPA rules required secondary and primary school systems to coordinate a management plan with a specific individual responsible for identifying asbestos-containing materials, evaluating hazards, and overseeing ultimate abatement of the hazard, for example, by removal, enclosure or encapsulation.

Effective July 10, 1995, OSHA rules went into effect that had first been published in February 1995 (60 Fed. Reg. 9624, Feb. 21, 1995). Employers must comply with the asbestos standard, 29 C.F.R. 1910.1001, especially:

1. engineering and work practice controls (1910.1001(f))
2. initial exposure monitoring ((d)(2))

3. regulated areas established after initial monitoring
4. respiratory protection (g)
5. construction plans for change rooms, showers, lavatories, and lunch-room facilities (i)
6. provide and ensure use of protective clothing and equipment if airborne exposures exceed PELs or may cause eye irritation (h)
7. limit means of sanding/stripping/buffing flooring (k)
8. post signs and labels warning of ACMs (j)
9. create medical surveillance not previously required (L)
10. written compliance programs required wherever exposure exceeds a PEL ((f)(2))

A typical academic or hospital building with insulation, ceiling, floor, or wall materials that are installed asbestos-containing materials (greater than 1% asbestos), is subject to these standards for general industry and construction activities. It is vital to note that "construction activities" that trigger tight OSHA regulations need not be the primary business activity of the employer. Activity as small as drilling a hole in wallboard that contains 1% or more of asbestos ingredients, or custodial work incidental to asbestos-related construction, is all covered.

Apart from construction activities, routine custodial work such as cleaning and washing around asbestos-containing materials is subject to general industry standards that set permissible exposure limits (PELs) at 0.1 asbestos fibers per cubic centimeter of air over an 8-hour time-weighted average. The 30-minute "excursion limit" is set at 1.0 fibers per cubic centimeter. In addition to requiring that exposures stay below PELs, revised standards also require implementation of other measures to hold exposures below the PELs.

Presuming that asbestos is present is very important. All known asbestos-containing materials must be identified before work begins, and the standards legally presume that asbestos exists in buildings built before 1981, unless proven otherwise, in the form of thermal system insulation, sprayed or troweled surfacing material, debris in work areas where either of those materials is present, and all asphalt and vinyl tile flooring material. *Proving* that no asbestos exists must be done by testing, not reliance on records about the construction.

4.2 TYPES OF WORK

The construction standard is divided into four hazard levels with appropriate control and training for each level. The levels range from highest (I) to lowest (IV) and include the following:

Class I is removal of thermal system insulation and sprayed on or troweled-on surfacing of asbestos-containing materials (ACM). Class II includes removal of other ACM such as floor tile, wall board, roofing and siding shingles, and construction mastics. Class III includes repair and maintenance operations where ACMs are likely to be disturbed, such as by drilling or cutting of small amounts. Class IV means maintenance and custodial activities that *contact* the materials, and cleanup of waste and debris containing ACM. OSHA requires an initial exposure assessment immediately before or at the initiation of every task covered by the standard unless the employer can demonstrate by a "negative exposure assessment" that exposures will be below the PELs.

Training requirements for the workers who will do Class I and II operations must be equivalent to that under the EPA Model Accreditation Plan for Asbestos Worker Training (40 C.F.R. Part 763, Subpart E, Appendix C). When contractors are bidding for work, advance notice must be given to them about ACM exposures in the work area.

The construction standard covers other asbestos operations (29 C.F.R. 1926.1101):

1. demolition or salvage on structures with ACMs
2. removal or encapsulation of ACMs
3. repair, construction, maintenance, alteration or renovation of structures, substrates or portions thereof that contain ACMs
4. installation of products containing asbestos
5. emergency cleanup of ACM spills
6. transportation, disposal, storage, containment of and housekeeping activities involving ACMs on the site or location where construction activities are performed

4.3 ASBESTOS CONTROL: INSTITUTIONAL REQUIREMENTS

The presumption that pre-1981 buildings used ACMs is rebutted by tests or by keeping on file written documentation that the material was no more than 1% asbestos. Also rebutting it is the finding that during any reasonably foreseeable use the material will not release asbestos fibers in excess of permissible exposure limits.

For post-1981 construction, the building operator should become aware of ACMs and identify any known ACMs, and then notify employees, tenants, and contractors.

4.4   ASBESTOS CONTROL: GUIDANCE
FOR BUILDING OWNERS

Owners should first identify materials that are over 1% asbestos, and record their location and type within the building. This record must be available for OSHA, tenants, employees or contractors to inspect. Then the suspect materials should be labeled, or if impractical to do so, a nearby sign is needed:

DANGER: CONTAINS ASBESTOS FIBERS. AVOID CREATING DUST. CANCER AND LUNG DISEASE HAZARD.

Tenants must be given information about the ACMs and it also must be given to contractors who do work in or around ACM. Signs must be posted around areas where ACM is located.

The broad reach of "asbestos hazard construction" includes driving nails into plaster or drywall that has ACMs, cutting plaster to service electrical outlets or lights, etc.

4.5   ASBESTOS CONTROL: CUSTODIAL
WORKER PROTECTION

Building managers have a choice. They can conduct air monitoring to determine if custodial worker exposure is below the permissible exposure level (PEL) of 0.1 fibers per cubic centimeter over an 8-hour period or 1.0 fibers over a 30-minute period. If the measurement shows less than the PEL, no action is needed except to be alert to changes in circumstances. Or managers may presume that the ACMs are present and give the required training and safety measures, including at least two hours of asbestos training annually; use of respirators equipped with high-efficiency particle absorbing filters; medical surveillance of workers; retention of health and exposure records for 30 years; bans on smoking in ACM exposure areas; restriction of access to ACM areas to designated workers; protective clothing must be given free to employees; and lunch areas must be placed outside the asbestos area.

These custodial workers' routine exposure triggers the above set of protective measures. Further and stronger protections apply to the maintenance and repair activities and to the cleanup of such activities.

4.6   ASBESTOS CONTROL: FLOORING

Vinyl, resilient, or asphalt flooring material that is presumed to contain asbestos
is subject to special housekeeping measures. No sanding or stripping of unwaxed
tile is allowed; dry buffing is allowed if the floor has at least three layers of wax;
and floor stripping must use low-abrasion pads at speeds below 300 rpm and wet
methods.

4.7   ASBESTOS CONTROL: EXPOSED
MAINTENANCE WORKERS

Routine maintenance that can disturb ACMs is classified under the OSHA
asbestos construction standard as Class III work. A job supervisor with adequate
training meeting OSHA requirements must oversee the work. Air monitoring of
airborne fibers is needed unless a negative exposure assessment has been
obtained. Cutting, drilling, and so on requires respirators, glove bags and boxes
or mini-enclosures. Training of 16 hours and 4 hours annual refresher is required.
Isolation of the job site by signs, barriers or negative pressure enclosures may be
required. Smoking and the use of certain saws and sweeping are prohibited. Wet
methods of handling ACMs are strongly preferred. Medical surveillance is
required for employees who perform this work with ACMs more than 30 days per
year.

4.8   ASBESTOS CONTROL: CLASS I AND II WORK

These are the classes of most serious exposure to asbestos. Class I is removal of
ACM in thermal system insulation or surfacing materials. Class II is removal of
all other ACMs. For these tasks the rules require:

1. a competent supervisor who has completed an EPA-required training
   course of 5 days with an annual 8-hour refresher
2. daily air monitoring or a presumption that PELs are exceeded
3. respirators must be used
4. smoking is prohibited in the work area
5. signs and barriers must be erected to warn others away
6. wet methods and other protective controls must be used

7. lunch areas must be separated from the ACM area
8. medical surveillance is needed for workers who perform ACM tasks 30 days or more per year
9. special negative pressure devices
10. certain prohibited work practices include dry sweeping

If PELs (0.1 fibers per cubic centimeter over 8 hours) are likely to be exceeded, the standard also requires air monitoring, supplier air respirators, protective clothing, decontamination procedures, mini-enclosures or negative pressure bags, and other safeguards.

Class II work has similar safeguards but is slightly less onerous in its requirements.

4.9   ASBESTOS CONTROL: CLASS III WORK

Maintenance work is in Class III. If airborne asbestos testing shows that PELs may be exceeded, for example, more than 0.1 fibers per cubic centimeter over an 8-hour period, then control measures similar to but less rigorous than the Class I and II steps above must be used. The OSHA asbestos standard should be consulted before the maintenance begins.

4.10   ASBESTOS CONTROL: CLASS IV WORK

The custodial workers that clean up after repair or removal of ACMs are engaged in Class IV work under the OSHA asbestos construction standard. A competent supervisor is needed and airborne fiber measurements must be conducted unless a negative exposure assessment is obtained. Other safeguard measures comparable to those of the Classes I, II, and III are also applicable, so the prudent safety manager will consult the OSHA asbestos construction standard before the custodial cleanup of an ACM work site begins.

4.11   CONFINED SPACE CONTROLS

In 1994, OSHA issued its long-awaited "Permit-Required Confined Spaces Standard" (29 C.F.R. § 1910.146). The standard requires that employers:

- identify all permit-required confined spaces in their facilities
- inform exposed employees and contractors of those spaces by signs or other effective means
- institute measures to prevent unauthorized entry into those spaces
- develop and implement a written permit space entry program to protect authorized entrants from permit space hazards

Before allowing entry into a permit space, the employer must, pursuant to the permit space entry program: (1) identify and evaluate the hazards of the space; (2) develop and implement appropriate entry, monitoring, and rescue procedures; (3) train and equip affected employees; and (4) inform contractors of the hazards identified and any procedures developed for dealing with them. The standard imposes less-stringent requirements in two situations: (1) where the atmospheric hazard is *controlled* by continuous ventilation; and (2) where the nonatmospheric hazard is *eliminated* prior to entry.

The standard requires practices and procedures to protect employees from the hazards of entry into a subcategory of "confined spaces" known as "permit-required confined spaces." The standard defines "confined space" to mean a space that: (1) is large enough and so configured that an individual can make a full body entry and perform assigned work; (2) has limited or restricted means for entry or exit (for example, tanks, vessels, silos, storage bins, hoppers, vaults, and pits); and (3) is not designed for continuous employee occupancy. A "permit-required confined space" or "permit space" is a confined space with one or more of the following characteristics: (1) contains or has a potential to contain a hazardous atmosphere; (2) contains material that may engulf an entrant; (3) has an internal configuration likely to trap or asphyxiate an entrant; or (4) contains any other recognized serious safety or health hazard (e.g., extreme temperatures).

An employer's confined-space permit program must address five general areas. The program must describe how the employer will: (1) identify and evaluate hazards; (2) develop and implement entry, monitoring, and rescue procedures; (3) train and equip affected employees; (4) assist any contractors hired for entry operations in complying with the standard by informing them of the identified hazards and any procedures developed for dealing with them; and (5) periodically review the effectiveness of the program.

4.12   CONFINED SPACES: MODEL PROGRAM

A sample program and required elements follows:

Introduction to Model Program

This written plan is designed to comply with the United States Occupational Safety and Health Administration (OSHA) 29 CFR 1910.146 "Permit Required Confined Spaces" requirements. This plan sets forth safety procedures to be followed in areas such as boilers, sewers, storage tanks, utility vaults, and manholes.

No one shall be allowed to enter a confined space without an authorized entry permit and training. This document will establish the procedures required to guarantee compliance with the regulation and maintain a safe working environment for all institutional staff, faculty, and students.

Definitions for Confined Space Plan

Acceptable Entry Condition: The condition that must exist in a permit space to allow entry and to ensure that employees involved with a permit-required confined-space entry can safely enter into a work within the space.

Attendant: A person who is assigned to monitor and protect the safety of authorized entrants to a confined space and must not perform other duties that would interfere with this.

Authorized Entrant: A person who is authorized by the supervisor to enter a confined space. He or she must be able to recognize signs and symptoms of exposure and know how to use any needed equipment, and how to communicate with the attendant as needed and alert them when some warning symptom or hazardous conditions becomes evident.

Confined Space: An enclosed space that has limited openings for entry and exit, poor ventilation, and is not designed for continuous worker occupancy.

Engulfment: The surrounding, capturing, or both, of a person by divided particulate matter or liquid.

Entry: The action by which a person passes through an opening into a permit-required confined space.

Entry Permit: A written or printed document that is provided by the employer to allow and control entry into a permit space.

Entry Supervisor: A person (such as the employer, foreman, or crew chief) who authorizes entry and prepares and signs the written entry permits after having made sure that all necessary tests have been conducted and that rescue services can be readily contacted.

Flammable Atmosphere: An atmospheric concentration of any substance for which a permissible exposure limit has been set by federal regulations.

Hazard Evaluation: A process to assess the severity of known, real, or potential hazards or all three, at or in confined space.

Hazardous Atmosphere: An atmosphere that may expose employees to the risk of death, incapacitation, impairment of ability to self-rescue, injury, or acute illness by reason of: oxygen deficiency or enrichment; flammability or explosivity; or toxicity.

Hot Work: Work within a confined space that produces arcs, sparks, flames, heat, or other sources of ignition.

Immediately Dangerous to Life or Health (IDLH): Any condition that poses an immediate or delayed threat to life or that would cause irreversible adverse health effects or that would interfere with an individual's ability to escape unaided from a permit space.

Inerting: The displacement of the atmosphere in a permit space by noncombustible gas (such as nitrogen) to such an extent that the resulting atmosphere is noncombustible.

Isolation: A process of physically interrupting, or disconnecting, or both, pipes, lines, and energy sources from the confined spaces.

Lockout/Tagout: The placement of a lock/tag on the energy-isolating device in accordance with an established procedure, indicating that the energy-isolating device shall not be operated until removal or the lock/tag in accordance with procedure.

Lower Explosive Limit (LEL): The lower limit of flammability of a gas or vapor at ordinary ambient temperatures expressed in percent of the gas or vapor in air by volume. This limit is assumed constant for temperatures up to 120°C (250°F). Above this it should be decreased by a factor of 0.7 because explosibility increases with higher temperatures.

LFL: Acronym for Lower Flammable Limit.

Oxygen Deficient Atmosphere: An atmosphere containing less than 19.5% oxygen by volume.

Oxygen Enriched Atmosphere: An atmosphere containing more than 23.5% oxygen by volume.

PEL: An acronym for the OSHA permissible exposure limit, which is the allowable air-contaminant level established by OSHA.

Qualified Person: A person who is designated by the Confined Space Program. This qualified person must meet the training, experience, knowledge, and education requirements.

Testing: A process by which the hazards that may confront entrants of a permit space are identified and evaluated. Testing includes specifying the tests that are to be performed in the permit space.

TLV: Acronym for threshold limit value, which is a time-weighted average concentration under which most people can work consistently for eight hours a day, day after day, with no harmful effects.

Toxic Atmosphere: An atmosphere containing a concentration of a substance above the published or otherwise known safe levels.

Upper Explosive Limit: The highest concentration (expressed in percent vapor or gas in the air by volume) of a substance that will burn or explode when an ignition source is present.

UFL: Acronym for upper flammable limit.

4.13   CONFINED SPACES: HAZARD IDENTIFICATION

A survey and inventory of confined spaces must be performed. The survey shall include scope, magnitude, possibility, and consequences of hazards such as, but not limited to, oxygen deficiency or enrichment atmosphere, flammable/explosive atmosphere, toxic atmosphere, biological, and mechanical hazards. The hazard evaluations should be updated annually and as deemed necessary due to changing conditions.

Atmospheric Hazards

A confined-space atmosphere can become extremely hazardous due to a lack of air movements. These hazardous atmospheres are classified as oxygen deficient/enriched, flammable/explosive, and/or toxic.

   The oxygen level in a confined space can decrease because of work such as welding, cutting, or brazing; or, it can decrease by certain chemical reactions (rusting) or through bacterial action (fermentation). The oxygen level also decreases if oxygen is displaced by another gas, such as carbon dioxide or nitrogen. This is known as an oxygen-deficient atmosphere. Another atmospheric hazard is an oxygen-enriched atmosphere. This occurs when the oxygen level exceeds 23% by volume. This condition represents a serious fire hazard. Flammable materials and hair will burn very rapidly in an oxygen-enriched atmosphere. Unattended or leaking oxygen lines or cylinders can increase the oxygen concentration in confined spaces to unsafe levels and should be recognized as potential hazards.

Flammable atmospheres are the product of gas, vapor, or dust in the proper mixture. Different gases have different flammable ranges. Before employees are permitted to enter any confined space containing flammable liquids, vapors, or gases, a qualified employee shall test the atmosphere within the space with a combustible gas meter to determine the concentration of flammable vapors.

Toxic atmospheres in a confined space can be the product of chemical products stored in the area, the work being performed in a confined space, and/or areas adjacent to the confined space. Therefore, before employees are permitted to enter any confined space containing liquids, vapors, gases, or solids of toxic, corrosive, or irritant nature, a Qualified Person must test for the presence of atmospheric contaminants. The two most common gases found in confined spaces are hydrogen sulfide (H2S) and carbon monoxide (CO). Both can be life threatening.

Hydrogen Sulfide is a toxic, colorless, combustible gas that is heavier than air. It is formed by the decomposition of organic plant and animal life by bacteria. Hydrogen sulfide poisons a person by building up in the bloodstream. This toxic gas paralyzes the nerve centers in the brain which control breathing. As a result, the lungs are unable to function and the individual is asphyxiated. Hydrogen sulfide can be found in oil and gas refining and production, sewers, pulp mills, and a variety of industrial processes. Hydrogen sulfide is easily detected by a strong rotten egg odor in low concentration. However, relying on this odor to warn of the presence of hydrogen sulfide can be very dangerous in certain conditions. High concentration can rapidly paralyze the sense of smell. Even low concentrations desensitize the olfactory nerves, after prolonged exposure, to the point that an individual may fail to smell the presence of the gas even if the concentration suddenly increases.

Carbon monoxide (CO) is a toxic, colorless, odorless, combustible gas that is slightly lighter than air. A by-product of combustion, it can be found in almost every industry. Carbon monoxide (CO) enters our bloodstream through the lungs. CO has an extreme affinity for the hemoglobin in our bloodstream at about 200–300 times that of oxygen. As a result, carbon monoxide quickly replaces oxygen in our bloodstream and causes asphyxiation.

Acceptable Limits

OSHA stipulates minimum "air levels" for acceptable occupancy of a confined space. These air levels are established by the American National Standard Institute (ANSI). Therefore, ANSI considers a confined-space atmosphere appropriate when:

- oxygen is between 19.5% and 23.5%
- flammability is less than 10% of the LEL or LFL
- toxicity is less than permissible exposure limits (PEL)

Atmospheric Testing

Depending on their weights, hazardous gases could be at the bottom, middle, or top of a given confined space. Some gases are heavier than air, others lighter, some the same weight. Therefore, the qualified person must test the atmosphere of a confined space at all levels (top, middle, bottom) with properly calibrated instruments to determine what gases are present. If testing reveals oxygen deficiency, or the presence of toxic gases or vapors, the confined space has to be ventilated and retested before any entry is permitted. When ventilation cannot be accomplished, employees must use respiratory protection if entry is required for emergency situations.

The testing and the calibration equipment used by the qualified person shall be approved by a recognized testing laboratory such as Underwriters' Laboratory (UL) or Factory Mutual Systems and be used and maintained in accordance with manufacturer's recommendation.

4.14   CONFINED SPACES: PHYSICAL HAZARDS

Other potential hazards can be encountered in confined spaces. They are as follows:

Extreme Heat or Cold

Temperature extremes could create health hazards and affect the amount of work that people can do and the manner in which they do it.

Noise

Excessive noise can damage hearing, and can also affect communication between the authorized entrants and the qualified person.

Slick/Wet Surface

Slips and falls can occur on a wet surface, causing injury or death to employees. Also, a wet surface will increase the likelihood and effect of electric shock in areas where electrical circuits, equipment, and tools are used. To prevent electric shocks, all electrical equipment must be grounded. The National Electric Code recommends a ground-fault circuit interrupter (GFCI) to sense leakage currents of a magnitude that could cause serious injury. A GFCI is a fast-acting electric circuit-interrupting device that is sensitive to very low levels of current flow to ground.

Falling Objects

Employees in confined spaces should be aware of the potential of falling objects, particularly in spaces with topside openings for entry, and where work is being done above the employees.

4.15   CONFINED SPACES: HAZARD CONTROL

Pre-entry Procedures

Confined spaces may contain atmospheres hazardous to life and health because of the presence of flammable, explosive, or toxic contaminants or because of deficiency or excess of oxygen. Therefore, the qualified person shall perform all the following steps:

   a. The space shall be emptied, flushed, steamed, or otherwise purged of flammable, injurious, or incapacitating substances to the extent necessary for entry without respiratory equipment.
   b. The air inside the confined space shall be tested with an appropriate device or method to determine whether dangerous air contaminants

and/or oxygen deficiency exists. The result of the test must be written down on the entry permit.

c. Oxygen can be depleted by using torches, flames, or other similar materials. Therefore, measures shall be taken to ensure that such devices do not lower the oxygen level below 19.5% and that all exhaust gases or fumes from operations such as welding are safely exhausted. To minimize the potential of oxygen depletion, mechanical devices such as blowers must be used to keep air movement in the area. Monitoring should be performed every 15 minutes. This is to verify the appropriate oxygen content in the area.

d. Forced-air ventilation (blowers) may also be used to maintain the 19.5% oxygen requirement. Blowers are necessary if the confined space or a portion of the confined space to be occupied is not naturally ventilated.

e. When working in confined spaces (tanks) which previously contained substances corrosive to the skin or substances which can be absorbed through the skin, employees shall be required to wear appropriate personal protective equipment. Therefore, the material-safety data sheet (MSDS) of the substances shall be reviewed. The MSDS is a summary of the important health, safety, and toxicological information on the chemical or the mixture ingredients. The MSDSs are located at the [central safety office] and in respective work areas.

f. Secure the area by using barricades and/or hazard warning tape.

Safe Entry

a. All provisions stated in the pre-entry procedures must be performed.

b. The qualified person must be positioned at the entry point of the confined space. Also, the qualified person must be aware of the nature of the work being performed and the possible hazards that might be encountered.

c. An effective means of communication between employees inside the confined space and the qualified person must be provided. Two-way radio, hand signals, or voice are acceptable forms of communication.

d. If a fire or emergency occurs in another portion of the plant, all persons in the confined space must be removed immediately. Reentry of the confined space is allowed once the emergency is over.

Permit System

A permit must be issued prior to the entry into a confined space. Permits will constitute written permission to perform designated work within a specified time period, and must be completed by the qualified person. The entry permit shall include:

- identification of the space
- purpose of the entry
- date and duration of the permit
- list of authorized entrants
- names of the current attendants and entry supervisor
- list of any hazards in the permit space
- acceptable entry conditions
- results of tests (initialed by the qualified person)
- rescue and emergency services available
- communication procedures for qualified person/entrants
- required equipment (e.g., respirators, communications device, alarms)
- types of safety equipment required

NO CONFINED SPACE SHALL BE ENTERED WITHOUT A SAFE ENTRY PERMIT!

Training

Employees in charge of permit-required confined space entries must be provided with full information and training. The training will be provided by the [Department title]. According to OSHA, the training shall cover the following topics:

a. Know the hazards which may be faced during entry.
b. Recognize the signs and symptoms of exposure to a hazard.
c. Understand the consequences of exposure to a hazard
d. Maintain contact with the standby qualified person
e. Notify the qualified person when the entrants self-initiate evacuation from the confined space.
f. Know what personal protective equipment is necessary.
g. Understand and use personal protective equipment, such as retrieval lines, respirators, or clothing, needed for safe entry and exit.

A *qualified person*'s responsibilities include the following:

- Plan each entry; including:
  Describe the work to be done.
  Evaluate the hazards of the space.
  Perform (or arrange for) atmospheric testing and monitoring.

Identify the employees involved.

Develop rescue plans.

- Make sure to complete, date, and sign the permit.
- Determine the need for specialized equipment. If ventilation equipment is necessary, make sure it works properly.
- Ensure atmospheric testing.
- Monitor the atmospheric testing equipment.
- Make sure all necessary procedures, practices, and equipment for safe entry are in effect.
- Check at "appropriate" intervals the atmospheric conditions in the confined space.
- Keep track of who is in the space at all times.
- Maintain continuous communication—visual, voice, hand signal, or by two-way radio—with the authorized entrant during the entry.
- Make sure that an entry permit shall be good for maximum of eight hours.
- Remain alert for early symptoms of danger within the space.
- Watch for hazards outside and inside the space.
- Maintain clear access to and from the space.
- Cancel the permit and terminate the work if the conditions are not acceptable, or if the permit limits expire.
- Cancel the permit and secure the space when the work is done.
- Remain at the entry point unless relieved by another qualified person.
- Determine if a written rescue plan is necessary for a particular confined space entry.
- Verify that emergency help is available and that the method of summoning help is operable.

An *authorized entrant*'s responsibilities include the following:

- Perform the assigned task.
- Review the permit before entry.
- Wear appropriate personal protective clothing.
- Use appropriate personal protective equipment.
- Use and attend to areas and personal monitoring equipment.
- Monitor physical reactions that could signal an unsafe condition.
- Maintain contact with the qualified person and respond to evacuation orders.
- If the entrant senses any reaction to the surroundings, he or she should signal the qualified person for help. If necessary, leave the confined space immediately.

Personal Protective Equipment (PPE)

A qualified person or the entry permit will indicate what, if any, personal protective equipment (PPE) is necessary prior to entering a confined space. Each department involved with this program must provide appropriate personal protective equipment.

Based on the hazard evaluation, a qualified person shall consider PPE for the following:

- gloves to protect against sharp edges, abrasion, punctures, chemicals, heat, cold electricity
- hard hat to protect against falling objects and structural configurations within the space, i.e. low ceilings or passageways
- eye and face protection to protect against irritation dusts, vapors, mists, abrasive particles, and flying objects
- hearing protection to protect against excessive noise within the space. Be sure that alarms are audible and communication with the qualified person is possible while using the hearing protection.
- clothing to protect against temperature extremes, moisture, chemicals, vapors, and toxic materials

Personal protective equipment DOES NOT ELIMINATE THE HAZARD. It provides protection from the hazard. To be effective, it must fit properly and be used in the manner for which it is designed.

Emergency Response

The qualified person is the first line of defense in an emergency. It is the qualified person's duty and responsibility to be able to recognize an emergency and to respond appropriately. The qualified person shall be in constant communication with authorized entrants in the confined space and shall have means such as two-way radio to summon help. This program does not provide an in-plant rescue team. Therefore, if there is an emergency which the qualified person recognizes as potentially life- or health-threatening, the qualified person SHALL NOT ENTER THE CONFINED SPACE, but must take the following steps:

a. The qualified person shall direct occupants to exit the confined space.
b. If an occupant has become unconscious or is otherwise unable to exit the confined space, the qualified person shall notify the Department of Public Safety, via the two-way radio or by phone. Information regarding

the emergency, location, potential hazards, and number of injured shall be provided to the Department of Public Safety at this time.

   c. The qualified person at the entry point must call 911 or other appropriate emergency number and direct them to the site of the emergency. A copy of confined-space locations and the written program will be provided to the emergency-response personnel.

   d. The qualified person shall stay at the site of the emergency and give assistance as directed by emergency response personnel.

In a situation which is not life- or health-threatening to the other occupants or the qualified person (such as an entrant with sprained ankle), the qualified person must take the following steps:

   a. The qualified person shall notify the emergency rescue group via two-way radio or by phone. Information regarding the emergency location, potential hazards, and number of injured should be provided at this time.

   b. The qualified person shall enter the confined space to render first aid or other assistance only if he/she is trained in such assistance and only if there is another trained person able to take over the duties of the qualified person.

Contractors

All departments that hire contractors to do work in confined spaces must require that they present proof of their Confined Space Program.

4.16 MANAGEMENT OF EXPOSURES TO LEAD

Lead exposure is regulated through several different laws and regulations covering food, air, housing, occupational exposure, disposal, and prevention of poisoning. For institutions that provide living quarters in older buildings for students and family housing for married students, this paint-contamination concern is especially important. The Lead-based Paint Poisoning Prevention Act, signed in 1971, established a proactive approach to eliminating lead poisoning. In 1973, the U.S. Consumer Product Safety Commission adopted a limit of 0.5% lead in any paint manufactured for use in or on housing. In 1976 the use of lead-based paint was prohibited for use on eating, drinking, and cooking utensils (42 U.S.C. 4831).

In 1978, the Consumer Product Safety Commission provided a more restrictive paint standard of 0.06% lead content. The limit is based on the percentage of solids in the paint or the paint film. In effect, this banned the use of lead-based paint in residential housing. Also, the use of lead-based paint on furniture, toys, and other items that could be used by children was banned (16 CFR 1303). For paint identified in either federally owned, subsidized, or Indian housing, tests are required to determine lead content and any paint found with greater than 0.06% lead per milligram per square centimeter must be abated (24 CFR 35).

In 1988, the Lead-based Paint Poisoning Prevention Act emphasized the need to protect the public from lead poisoning by removing lead-based paint from their homes.

Title X of the Community Housing and Development Act (42 U.S.C. 4822), also known as the Residential Lead-based Paint Hazard Reduction Act of 1992, was significantly more specific than previous legislation in preventing lead poisoning. In addition to protecting the public from the hazards of lead-based paint exposure, the potential hazards of exposure to lead dust is addressed. Title X will have a significant impact on lead-based paint inspections, assessments, and abatement in residential, commercial, industrial, and publicly funded housing. Title X details provisions for training, licensing, and certification of abatement contractors. Housing is eveluted for the level of hazard from the lead paint and dust. The hazard is then abated according to the eveluation, whether temporary or permanent.

EPA may set environmental limits for lead in dust and soil, but until it does so, the Department of Housing and Urban Development (HUD) will be taking the primary role in this area. Although HUD standards are not enforceable by law in most institutional settings, they do provide a standard that can be used as a yardstick when dealing with lead evaluation and abatement. HUD publications can be requested from HUD, P.O. Box 6091, Rockville, Maryland 20850 or (800)245-2691.

The Resource Conservation and Recovery Act (40 CFR 261.24) does consider any waste with more than 0.5 milligrams of lead per liter as a hazardous waste. Hazardous waste containing lead must be manifested by a generator, carried by an EPA-registered transporter, and disposed of in an EPA-approved treatment, storage, and disposal facility. Whether lead is a hazardous waste is determined by the use of a toxicity characteristic leaching procedure (TCLP). This test is based on the percentage of lead that leaches out of a sample of the test material under laboratory conditions that mimic the release of toxic leachates from similar waste buried in a landfill over a long period of time.

The maximum amount of lead allowed in drinking water is 15 parts per billion. This limit is set through the Safe Drinking Water Act of 1986 (42 U.S.C. 300a) which is addressed in text Chapter 12. Drinking-water fountains that have lead-lined tanks or could cause lead poisoning were prohibited from sale in 1988. By 1990, each state was to have established a program for dealing with this problem. Presently, bottled water is regulated to maintain lead content below 0.005 milligrams of lead per liter by the Food and Drug Administration.

The OSHA lead standard for general industry (29 CFR 1910.1025) was first published in 1978. The standard set a permissible exposure limit of 50 micrograms of airborne lead per cubic meter of air. The standard required the use of feasible engineering and work-practice controls. The standard applies to all employers covered by OSHA, except construction and agricultural operations.

The construction industry is covered by the Construction Industry Interim Lead Standard (29 CFR 1926.62). Construction work is defined as work involving construction, renovation, alteration, repair, painting, and decorating when lead is present.

Exposure to airborne concentrations of lead above or equal to an action level of 30 mg/m3 calculated as an eight-hour time weighted average (TWA) for employee exposure, without regard to the use of respirators, triggers periodic exposure monitoring, biologic monitoring, and employee training requirements whenever exposure measurements equal or exceed the action level. If employees are exposed above the action level for more than 30 days per year, the employer must provide a periodic medical-surveillance program for those affected employees.

Employers are required to conduct initial exposure assessments to determine if any employee is exposed to lead at or above the action level of 30 mg/m^3 as an eight-hour TWA. Instead of basing this assessment exclusively on air-monitoring data, objective information demonstrating that a specific process, operation, or activity cannot result in employee exposure to lead at or above the action level during expected conditions of use or handling can be used.

Historical measurements of airborne lead concentrations can be used to satisfy the initial exposure-assessment requirement if the measurements have been taken within the previous 12 months. The previous workplace conditions should closely resemble those conditions used in the current operations.

The employer must ensure that no employee is exposed to airborne concentrations of lead in excess of the permissible exposure limit (PEL) of 50 mg/m^3 as an eight-hour TWA.

If the initial determination or subsequent determination reveals employee exposure to be at or below the PEL, monitoring must be performed at least every six months. If the initial determination reveals that employee exposure is above the PEL monitoring must be performed quarterly.

Also, certain tasks or operations involving lead are presumed to expose employees to levels greater than the PEL and interim protective measures should be implemented by the employer until the employer demonstrates exposure levels lower than regulatory limits.

Employees must be given written notification of the results of their exposure assessments within five working days.

4.17   MANAGEMENT OF EXPOSURES TO LEAD: COMPLIANCE

Employers must institute engineering and work-practice controls to reduce exposures to or below the PEL. If feasible controls are not adequate, the employer is required to provide appropriate respiratory protection.

When administrative controls are used as a means of reducing employees' exposure to lead, a job-rotation schedule should be implemented and employees should follow appropriate work practices.

The employer must develop and implement a written compliance plan prior to starting a job where employee exposure to lead, without respect to respiratory protection, will be in excess of the PEL. The plan must be reviewed and updated at least every six months.

The compliance program must provide for frequent and regular inspections of job sites, materials, and equipment to be made by a competent person. A competent person is defined as one who is capable of identifying existing and predictable lead hazards in the surroundings or working conditions. The competent person should have authorization to take prompt corrective measures to hazardous conditions.

4.18   CONSTRUCTION SAFETY: INTRODUCTION

Construction safety tends to overlooked in most insititutional settings. The reason is not because the safety component is not performing their function, but because of limited resources. Usually construction jobs are contracted to outside entities. As a result, contractors perform most construction jobs at universities, hospitals, and schools. When an institution's safety resources are stretched, it becomes very convenient to rely on the contractor's safety program to perform that function. Each contractor's safety program should be requested, compiled, evaluated, and retained on file, before work is to be done on site. Periodic evaluations should be conducted of the on-site work of these contractors.

4.19   CONTRACTOR SAFETY: CONTRACTUAL REQUIREMENT SUGGESTIONS

Contractors should be required to attest in the contract that they will comply with all OSHA, EPA, DOT, and other relevant local laws and regulations.

Prior to starting a job, all contractors should be expected to be aware of all personal protective equipment (PPE) required to perform the contracted job safely. All PPE should be either NIOSH- or MSHA-approved equipment.

Contractors using respirators shall be able to demonstrate compliance with the "Respiratory Protection" provisions of 29 CFR 1910.134 and 1926.58.

Contractors should be required to maintain readily available material-safety data sheets for all chemical substances that they or their employees could be exposed to in accordance with the Hazard Communitcation Standard (29 CFR 1910.1200).

Contractors shall erect and maintain appropriate access-restricting barriers and warning signs to prohibit entry into construction work sites by unauthorized personnel.

All pedestrian areas must be protected from activity that might result in debris, tools, or other materials falling onto pedestrian-trafficked areas. Well-constructed temporary barriers or restrictions in conjunction with warning signs should be used to protect these areas.

All energy sources that are locked or tagged out in accordance with 29 CFR 1910.146 should be properly restored prior to using the equipment or at the completion of the contractor's work.

Contractors who have to enter a confined space in the course of their work must comply with the provisions of the Confined Space Standard, 29 CFR 1910.146.

Contractors who are performing operations that could result in environmental indoor contamination (particulate, vapor, gas contamination) should give prior notification to the appropriate authority (insert name, phone number) having jurisdiction in such matters.

Ropes, hoists, and pulleys used for lifting equipment and materials must be rated for the appropriate capacity.

Welding operations require prior written notice to the insitutional fire marshal (insert name, phone number). Appropriate PPE, welding screens where others could be impacted, fire extinguishers, and a fire watch as necessary should be provided by the contractor.

If asbestos or other hazardous material is encountered during the course of a project, then the individual responsible (insert name, phone number) with duties in that area should be notified immediately.

Contractors will conform to standard life-safety practices by not blocking approved fire lanes, means of egress, fire extinguisher, or other emergency equipment with vehicles, construction materials, debris, equipment, or other encumbrance. Entrances to mechanical and electrical rooms and vaults will left clear.

Fire protection on the job site is the responsiblity of the contractor. Appropriate fire extinguishers are required in the work area and necessary fire watches should be provided by the contractor.

Construction vehicles will have operable back-up alarm devices or some other means of warning.

Visual inspections of construction sites should be conducted periodically. Observations and recommendations should be documented and distributed to the affected parties.

4.20   CONTRACTOR SAFETY: OSHA REQUIREMENTS

Section 1926.28 imposes a general duty on employers to provide employees, and require them to wear, "appropriate personal protective equipment in all operations where there is an exposure to hazardous conditions or where this part indicates the need for using such equipment to reduce the hazards to the employees."

4.21   APPLICATION OF FALL-PROTECTION STANDARDS

One area of recent OSHA concern affecting the construction industry is fall protection. The following case details how important it is to have all safety elements at a construction site effectively in place.

Construction work has always been known as a dangerous trade. Steel erection has unique hazards, particularly from falls due to the heights at which such work must often be done. To mitigate hazards involved in steel erection, OSHA promulgated a specific standard as SubPart R of the Constructions Industry standards (C.F.R. § 1926.750). Additional provisions to protect employees from falls are found in § 1926.28(a) requiring personal protective equipment to guard against such fall hazards (CFR § 1926.753). These were updated in 1994 (OSHA 1994), requiring perimeter safety nets when other forms of fall protection are impractical. The case law illustrates (*Peterson Brothers v Reich* 1994) how these standards are integrated into a comprehensive system of fall protection in the construction industry.

Section 1926.750 requires temporary floors to be installed within two stories or 30 feet below any tier of beams on which work is being done, or where these are not feasible, a safety net on the interior of a building structure when the potential fall distance is two floors or 25 feet. At the floor periphery, a safety railing of 1/2-inch wire rope, approximately 42 inches high, must be installed. Section 1926.105 requires that safety nets be provided when workplaces are more than 25 feet above the ground or water surface, or other surfaces where the use of ladders, scaffolds, catch platforms, temporary floors, safety lines, or safety belts is impractical.

Contractors should be required to comply with the OSHA fall-protection requirements for walking and working surfaces involved in construction work and other standards that require protection against fall hazards. These standards include Scaffolds, Subpart L; Personal Protective Equipment § 1926.95; Safety Nets § 1926.105; Cranes and Derricks, Subpart N; Steel Erection Subpart R; Underground Construction, Caissons, Cofferdams, and Compressed Air, Subpart S; Power Transmission and Distribution, Subpart V; and Stairways and Ladders, Subpart X.

In the 1990s, OSHA has issued several memoranda clarifying its position on these requirements. OSHA views the Steel Erection Standard, the Safety-Net Standard, and the Personal Protective Equipment Standard at § 1926.28 as an integrated system designed to address all fall hazards in steel erection. Falls to the interior of buildings must be prevented by providing flooring under § 1926.750. Section 1926.753 requires safety nets to protect against perimeter falls, and 1926.28 requires personal fall protection when the protection provided by §§ 1926.750 and 1926.105 is inadequate.

Additionally, OSHA has a policy to cite employers under the General Duty Clause when employees are exposed to falls of less than 30 feet. Sections 750 and 105 cited above only address falls of 25 feet or greater. OSHA will seek to show that an employer is aware that a fall of less than 30 feet can cause serious injury.

Two disputed cases show how fall issues are debated. For example, in the *Peterson* case, the company was erecting the structural steel framework at a site in Austin, Texas, when an employee, performing the job of "connector," was knocked off a beam, falling 70 feet to his death. The connectors install the first two bolts on horizontal beams after raising the columns for the structural steel frame of a building, always at the top of a building frame as it is being erected. While installing the beams, the Peterson Brothers' connectors wore safety belts, but no safety lines or lanyards were attached, and the company did not install safety nets at the perimeter of the building. As a result, at the inspection following the fatality, the company was cited for failing to install the nets as required by § 1926.105(a).

The employer argued that it could not be cited under this standard because it did not know that the section applied to its operations. If a reasonable employer in Peterson Brothers' position would not have known that the section applied, the employer cannot be cited because the employer would not have adequate notice of its specific legal obligations. This concept of "notice" is part of every citizen's right to due process guaranteed by the Fifth Amendment of the United States Constitution.

OSHA argued that the employer knew that the standard applied to steel erection. In an earlier case, Peterson had a similar citation vacated by an administrative law judge (ALJ) on the basis of two prior Occupational Safety and Health Review Commission (OSHRC) cases, which held that only Subpart R, the Steel Erection Standard, applied. Those cases were reversed on appeal, as noted in the

ALJ's decision, but the Court acknowledged that the case law was unclear. Peterson, it said, should have known that its reliance on the OSHRC decisions was misplaced. By the time of the citation, at least four circuit-court decisions had held that the Steel Erection Standard in Subpart R did not preempt the safety-net requirement because the particular hazard of a fall at the perimeter beam was not addressed in the Steel Erection Standard.

Peterson also argued that it was not technically feasible to comply because safety nets must be supported from above. Because the connectors' job was to erect the supports at the level at which the nets would be attached, there was nothing to support the nets within 25 feet of the height at which they were working. However, the Court upheld the Occupational Safety and Health Review Commission (OSHRC) decision that Peterson had to comply to the extent possible. Limited compliance is required where it provides some protection. At least, the Court said, nets should have been installed three stories below.

Peterson Brothers was therefore held to have violated the standard by not installing safety nets on the perimeter of the building during erection of the structural steel frame. Moreover, the implication of the decision is clear that the nets would have to be raised so that they are no farther than three stories below the height where the Connectors are working at any given time. If it is technically feasible for an employer to reduce the free-fall height to 25 feet or less, it must do so to comply with the specific provisions of the standards.

Under the General Duty Clause and the case law (*General Dynamics*), employers must take additional measures to mitigate a risk that the employer knows is not adequately addressed by a specific standard. In *General Dynamics*, the court said:

> If an employer knows a particular safety standard is inadequate to protect his workers against the specific hazard it is intended to address, or that the conditions in his place of employment are such that the safety standard will not adequately deal with the hazards to which his employees are exposed, he has a duty under section 5(a)(1) to take whatever measures may be required by the Act, over and above those mandated by the safety standard, to safeguard his workers. In sum, if an employer knows that a specific standard will not protect his workers against a particular hazard, his duty under section 5(a)(1) will not be discharged no matter how faithfully he observes that standard.

Thus, compliance with a specific OSHA standard, no matter how precise or complete, will not absolve an employer from a greater duty if additional steps are necessary to protect employees and such steps are feasible.

4.22   ELECTRICAL SAFETY—LOCKOUT/TAGOUT

Universities, hospitals, and schools that must comply with Federal Occupational Safety and Health Act standards must establish a lockout/tagout program to comply with Hazardous Energy Source Control: Subpart J (CFR 1910.147). The lockout/tagout program establishes safety requirements to prevent accidents from hazardous-energy sources by means of lockout/tagout procedures during the servicing and maintenance of machines and equipment. Hazardous-energy sources could include electrical power, stored mechanical energy (springs, gravity), pressure (air, oil, water, or other fluid), or stored electrical energy (transformers, capacitors).

Construction and agriculture operations are not covered by this standard.

This standard establishes minimum performance requirements for the control of such hazardous-energy sources and requires employers to develop and use an energy-control program. The program should include step-by-step procedures for the use of appropriate lockout or tagout devices to prevent unplanned energization, start-up, or release of stored energy which could cause injury to employees. Procedures must include preparation for equipment shutdown, isolation, use of lockout/tagout procedures, release of stored energy, and verification of isolation.

Case Study of Lockout Program Needs

OSHA is scrutinizing the elements of an employer's lockout/tagout program very closely as demonstrated by the *General Motors* case.

On April 4, 1991, Donald Smith, a skilled trades millwright at General Motors Corporation's (GM) Oklahoma City vehicle assembly plant, was killed while attempting to replace certain parts on a lift unit. As a result of the ensuing OSHA investigation, GM was accused of 57 violations of the Lockout/Tagout (LOTO) Standard (C.F.R. § 1910.147). The LOTO Standard requires that employers establish programs and procedures to control the release of hazardous energy during service and maintenance of machinery. The administrative law judge (ALJ) found that GM's LOTO program, which had been developed in conjunction with the United Auto Workers (UAW) under the National Joint Health and Safety Committee, failed to meet the requirements of the standard in that it did not "detail the specific steps needed to safely de-energize and lock-out particular equipment." In addition, the ALJ concluded that GM also failed to audit the performance of authorized employees in following the program and procedures, and that employees were not adequately trained in LOTO procedures. The majority of the citations were affirmed, as were penalties totalling $1,945,000. Both GM and OSHA appealed the decision to the Occupational Safety and Health Review Commission (OSHRC).

Donald Smith had been employed at the Oklahoma City plant for ten years. Although he had received the UAW-GM LOTO training in 1986, he was not familiar with the equipment he was to service, having only recently transferred into the area. He knew the plant procedure required him to shut off power and "dump the air" before performing his repair job. He sought the assistance of a more experienced individual and two supervisors in the area before beginning. The ALJ found that the lift activated when Smith was leaning into the lift area and struck him in the head and killed him. What caused the lift to cycle was not determined.

Based on the evidence produced at the trial, the ALJ concluded that the LOTO program was deficient. The program had been in place since the plant opened in 1979, and a written procedure described in a nine-page pamphlet established a general method for controlling hazardous energy during servicing of equipment in the plant. In 1986, this was supplemented by a 116-page training manual, and training using this manual was conducted at Oklahoma City between 1986 and 1989 by plant personnel (who had attended a special course at a New Jersey training center to become trainers). The course presented at the plant included hands-on lockout of some equipment in the classroom during the eight-hour course. Although the training apparently was provided, many of the employees interviewed by OSHA could not remember the training, and a number who were determined by the ALJ to be authorized employees requiring training under the standard had not received it.

The ALJ also determined that, in many cases and often in view of supervisors, the LOTO procedures were not followed and employees worked on unfamiliar equipment without receiving specific instructions on de-energizing that equipment. GM argued that the general procedure was adequate, if followed, to control energy on any equipment in the facility, and that the standard did not require separate procedures on each piece of equipment. The ALJ rejected this contention, holding that the standard required equipment-specific procedures and accepted the testimony of OSHA's expert on the standard that the complexity of the equipment determines the detail and specificity of the procedure that must be developed. Based on the lack of specific instructions on specific equipment, the ALJ concluded that the employees were not adequately trained.

In appealing the decision, GM argued that its program was adequate, the general procedure was sufficient to satisfy the standard, and the employees were adequately trained. GM's LOTO program, it argued, had been accepted by ALJs in other similar cases. Second, based in part on decisions in other related cases and on the preamble to the final rule, GM asserted that general procedures are sufficient so long as the types of controls are sufficiently similar among the difference machines covered.

With respect to the training requirement, GM argued that employees had been trained and that the ALJ incorrectly concluded that a training violation had been established if OSHA could demonstrate (1) that exposure to hazardous energy occurred, and (2) a failure to follow lockout procedures. Moreover, GM argued that, if the violations were affirmed, they could not be characterized as willful or egregious in light of GM's extensive program.

The GM case is important to hospitals and educational institutions because it illustrates OSHA's interpretation regarding what kinds of written program and procedures satisfy the LOTO standard. Employers should evaluate their programs and procedures in the context of the kinds of equipment being serviced by employees. Moreover, documentation of training should be reviewed to assure that the employer can demonstrate that training occurred, and if possible, that it was adequate.

Successful Internal Procedures

Program procedures for lockout/tagout should include the following:

- an evaluation of each operation requiring lockout/tagout
- specific procedures for shutting down, isolating, blocking, and securing machines or equipment to control hazardous energy
- criteria and responsibilty for the placement, transfer, and removal of lockout or tagout devices
- requirements for evaluating a machine or equipment to determine and verify the effectiveness of lockout/tagout devices, and other energy-control measures
- periodic (at least annual) inspections of the program should be part of the energy control program. This should be done by someone other than the individual performing the lockout/tagout procedures

The employer must ensure that before any employee performs any servicing or maintenance on a machine or equipment, the machine or equipment is isolated and rendered inoperative. When possible, lockout should be used instead of tagout. If an energy-isolating device cannot be locked out, the employer must use a tagout system.

When a tagout device is used on an energy-isolating device, the employer must demonstrate that the tagout device will provide a level of safety that is equivalent to that of a lockout system and the device must be attached at the same location that the lockout device would have been attached.

OSHA defines equipment that is "capable of being locked out" as that designed with a hasp or other fastener that a lock can go through or be affixed on, or that which has a locking mechanism built into it.

A lock must be used on equipment, if doing so will not require the employer to dismantle, rebuild, replace, or permanently alter the equipment's controller, the mechanism that energizes the equipment. For example, some controllers are not designed to be locked, but they can be secured with chains, wedges, or other similar devices which can be locked.

The standard requires that when switches, circuit breakers, or other such devices are installed in a single cabinet or box, employers tag the specific switch or device, not the cabinet or box.

Also, since January 1990, all major replacement, repair, renovation, or modification of machines or equipment, and all new machine or equipment installations have been required to be designed to accept lockout devices.

All devices used for isolation in a lockout/tagout program should be provided by the employer. Lockout/tagout devices must not be used for other purposes.

Lockout/tagout devices also must:

- be durable and capable of withstanding exposure to the environmental conditions to which they are subjected
- be standardized within the facility in either color, shape, or size
- be substantial enough to prevent removal without the use of excessive force or unusual techniques. Tagout devices and the means to attach them must be substantial enough to prevent inadvertent or accidental removal.
- identify the employee applying the devices. Tagout devices must warn against hazardous conditions if the machine or equipment is energized and must include appropriate warnings such as: Do Not Energize, Do Not Open, Do Not Close, and Do Not Operate.

Tagout devices must also:

- be able to withstand weather conditions
- not deteriorate when used in corrosive environments
- use standardized print and format

The standard also requires training for authorized and affected employees. Employee classifications fall into one of three classifications:

- *Authorized Employees:* persons actually responsible for physically locking or tagging out equipment
- *Affected Employees:* persons who could be affected by the hazardous energy source (operated affected equipment or work in the area)
- *Other Employees:* persons who may not be directly affected by the lockout/tagout process but who needs to be able to recognize when a lockout/tagout operation is in progress

Energy-Control Program

The employer must establish written procedures for a hazardous energy-control program that includes for the control of potentially hazardous energy when employees are engaged in maintenance and/or servicing activities.

Training

Employees must be trained to understand the lockout/tagout program and have the knowledge and skills needed for safe application, usage, and removal of energy-control mechanisms. The employer must have written certification that each employee has received training.

The training program for authorized employees (those who are charged with the responsibility for implementing the lockout/tagout control procedures and performing the service and maintenance) must cover at a minimum, the following areas:

1. details about the type and magnitude of the hazardous energy sources present in the workplace
2. the methods and means necessary to isolate and control those energy sources

By contrast, affected employees (usually the machine operators or users) and all other employees need only be able to:

1. recognize when the control procedure is being implemented
2. understand the purpose of the procedure and the importance of not attempting to start up or use the equipment that has been locked or tagged out

Because an affected employee is not one who is performing the servicing or maintenance, that employee's responsibilities under the energy-control program are simple:

Whenever there is a lockout or tagout device in place on an energy-isolating device, the affected employee leaves it alone and does not attempt to operate the equipment.

Retraining must be provided, as required, whenever there is a change in job assignments, a change in machines, equipment, or processes that present a new hazard, or a change in energy-control procedures. Additional retraining must be conducted whenever a periodic inspection reveals, or whenever the employer has reason to believe, that there are deviations from or inadequacies in the employee's knowledge or use of the energy-control procedure.

4.23   INDOOR ENVIRONMENTAL QUALITY

Concerns over the effects of airborne pollutants and their potential impact on human health have propelled the issue of indoor air quality to the forefront of workplace problems faced by building owners, employers, facility engineers, building managers, and employees in both the public and the private sectors.

The problem of poor air quality has been around for hundreds (thousands?) of years. The first caveman realized early on that barbecuing in the home cave resulted in very serious consequences. The early Egyptians believed exposure to silicate dust caused respiratory disease in stone masons. Throughout history, it has been recognized that individuals working in certain trades have experienced adverse health effects due to exposure to their environments. A treatise on occupational diseases by Bernardo Ramazzini, an Italian physician, was published in 1700. Not until the last few decades have serious efforts been directed toward improving air quality.

The air-pollution tragedies of the Meuse Valley, Belgium, in 1930 (Salvaggio 1990), Donora, Pennsylvania in 1948 (Schrenk, Heimann, and Clayton 1948), and London, England, in 1952 (Her Majesty's Stationary Office 1954) resulted from high levels of airborne pollutants concentrated by climatic conditions. The realization that this threat to public health could lead to increased morbidity and mortality has resulted in regulatory efforts to reduce this menace. Generally these regulatory efforts have targeted specific air pollutants. In the United States, federal air-pollution efforts began in 1955 with Clean Air Legislation, followed by the Clean Air Act of 1963, the Air Quality Act of 1967, and the 1970, 1977, and 1990 Clean Air Act Amendments. These acts consist of primary standards to protect health and secondary standards to protect the public. These regulatory actions, along with technical advances in air-pollution control, have succeeded in improving outdoor air quality.

However, the success with ambient air quality has resulted in a new awareness of the potential adverse health effects due to poor indoor air quality (IAQ). In the 1980s, a more educated general population began to realize that indoor air pollution could be just as serious a problem as outdoor air pollution. Postwar building booms with new improved materials, the tight buildings of the energy-efficient 1970s, the increase in technology available to both the offices and homes of the 1980s, and the simple fact that people spend most of their time indoors (Robinson 1977) (studies of human activities indicate that on an average, people spend approximately 22 hours per day indoors (Moschandreas 1981)) have all led to indoor air pollution becoming an important public-health issue.

Research has shown that indoor air quality can be affected by both the pollutants found in the ambient air and from those that originate within the structure (furnishings, operations conducted inside, and the structure itself). Conditions that can exacerbate problems with pollutant levels range from poor ventilation rates to energy-efficient (tight) buildings.

A typical institution has many conditions that contribute to the development of poor IAQ: combustion sources (cars, buses, incinerators, diesel-powered equipment, etc.), printing operations, and activities ranging from operations conducted in simple chemical lab classes to complex research projects. Typically these activities are performed in or near multiuse buildings (classrooms, offices, and patient-care areas).

IAQ problems manifest themselves often as nonspecific symptoms rather than clearly defined illnesses. Symptoms commonly attributed to IAQ problems include: headaches; fatigue; shortness of breath; sinus congestion; cough; sneezing; eye, nose, and throat irritation; skin irritation; dizziness; and nausea. However, all these symptoms can be caused by other factors and are not necessarily due to air-quality deficiencies. Environmental stressors such as improper lighting, noise, vibration, overcrowding, ergonomic stressors, and job-related psychosocial problems (such as job stress) can produce symptoms similar to those associated with poor air quality (EPA 1991).

The first IAQ conference, "Improving Indoor Air Quality," was held at South Berwick, Maine, in 1972. Two of the main issues discussed were that the internal generation of pollutants from sources such as gas stoves was a more important problem than previously suspected and that a doctor reported that some of his sensitive patients were apparently responding adversely to unidentified compounds generated indoors (Engineering Foundation Conference 1972).

4.24 SICK-BUILDING-SYNDROME (SBS) ALLEGATIONS

Complaints by a significant percentage of a building's worker or resident population about acute discomfort (e.g., fatigue, congestion, eye irritation, headaches), with immediate relief on leaving the building, is suggestive of what has been called in the press the "sick"-building syndrome. Causes of the problem are not recognizable, and successful mitigation is elusive (Walkinshaw 1988).

This term is used to describe cases in which building occupants experience acute health and comfort effects that are apparently linked to the time they spend in the building, but in which no specific illness or cause can be identified. Many different symptoms have been associated with SBS, including respiratory complaints, irritation, and fatigue. The problem may be caused by any or all of the following: the combined effects of multiple pollutants at low concentrations, other environmental stressors (e.g. overheating, poor lighting, noise), ergonomic stressors, job-related psychosocial stressors (e.g., overcrowding, labor-management problems), or unknown factors (EPA 1991).

The potential defendants in an SBS case include building owners, building managers, real-estate developers, architects, engineers, general contractors, heating, ventilation and air-conditioning contractors, building product manufacturers, indoor air-quality consultants, leasing agents, and energy management/ventilation consultants. Among these, the targets will be building owners, who have a duty to provide safe premises; the building manager, who must ensure the building systems operate properly; and all entities that participate in the design, construction, and installation of the building's ventilation system. Most institutions have many of the potential defendants on staff and could be vulnerable for SBS lawsuits.

There have been several lawsuits initiated in response to SBS. One of the first (*Buckley* v *Kruger-Benson-Ziemer* 1987) illustrated how far such a suit could extend. A computer programmer sued nine named and 280 unnamed defendants for injuries allegedly sustained from exposure to indoor air pollutants in his workplace. The defendants included all those listed above and more. The complaint argued that the plaintiff should have been warned by the defendants of poor ventilation and dangerous chemicals and toxins in the air, carpet, tile, and office machinery. The case reportedly settled for $662,500.

Illness allegations can be defined as: diseases or infirmities resulting from exposure to indoor air contaminants, usually characterized by clinical signs (e.g., blood serology, fever, infection, tissue deterioration), a low attack rate, and prolonged recovery times after leaving the building. Examples include humidifier fever from fungi and bacteria; legionellosis from bacteria; and toxicity from carbon monoxide, radon, asbestos, and mycotoxins. Successful mitigation requires removal of the pollutant source rather than better ventilation (Walkinsaw 1988).

Building-related illnesses as a category refer to illnesses brought on by exposure to the building air, where symptoms of diagnosable illness are identified (e.g., certain allergies or infections) and can be directly attributed to environmental agents in the air (EPA 1991).

The institution that expects to be protected from indoor air-quality liability claims by workers' compensation laws can find them to be circumvented because of an institution's peculiar set of circumstances. Quite often the institutional defendant is not only the employer but the owner and building manager, such as the university physical plant office and the dormitory manager.

Some prudent measures that can be taken to prevent or minimize indoor air liability are as follows:

- New construction and renovations should use products that have been tested in accordance with established scientific protocols and that have been shown not to adversely affect indoor air quality.
- Employees should be informed about SBS and building-related illnesses.
- A mechanism for reporting indoor air-quality complaints should be put into place.

- Employ preemptive measures when possible. Preventive maintenance and responding to potential problems before they reach the complaint stage can limit liability.
- Baseline testing of basic IAQ parameters should be conducted when possible.
- If complaints or problems are discovered, a plan to address the situation should be formulated, implemented, and documented.
- Activities that could affect IAQ should be recognized and their potential impact should be evaluated prior to initiation. These range from normal preventive maintenance on HVAC systems to renovation and construction operations. Although emergency situations that could affect IAQ can only be evaluated for their potential impact during the emergency, the mechanism for communication of the potential effects of the emergency can be put into place prior to the emergency taking place.
- Industry standards should be adhered to when designing systems that could affect IAQ. Other air-quality standards (ASHRAE 55-1981; ASHRAE 62-1989; ACGIH 1985; Ontario Ministry of Labour 1988; WHO 1977) are frequently discussed when applied to nonindustrial indoor environments.
- Occupational Safety and Health Administration (OSHA) permissible exposure limits (PELs) and American Conference of Governmental Industrial Hygienist (ACGIH) threshold limit values (TLVs®) were designed for healthy working adults in industrial settings and their applicability to nonindustrial environments is a matter of much discussion. They may not adequately reflect the health realities of today's institutional population (patients, students, office workers). Care should be taken when these standards are applied to a nonindustrial type environment.

4.25   A PROACTIVE APPROACH TO IAQ MANAGEMENT

The Environmental Protection Agency (EPA) and the National Institute of Occupational Safety and Health (NIOSH) have developed guidelines for building owners to follow when investigating IAQ complaints (EPA 1991). This publication can be used for general guidelines during the investigation. The "Note to Building Owners and Facilities Managers" in the introduction to the guidance manual points out that one of the most important goals to keep in mind is to establish a process that encourages an active exchange of information. There are many instances of IAQ investigations where open lines of communication were not maintained and the investigations did not satisfy the building occupants. The resultant atmosphere of distrust has escalated many simple IAQ investigations into complex scenarios involving the inefficient use of large amounts of resources and manpower.

The single most important response action during an indoor air-quality investigation is to maintain open lines of communication among all interested parties. If lines of communication are not kept open and updates are not forthcoming, it becomes very difficult, if not impossible, to resolve the complaint to everyone's satisfaction. In keeping with maintaining open lines of communication, one of the first issues to be resolved is what would constitute a successful end to an IAQ investigation. Investigators and building occupants need to be clear about what will constitute a successful resolution to the indoor air-quality problem.

An active communication network will also minimize complaints from building occupants. The building occupants need to be made aware in advance of routine or emergency procedures that could affect IAQ. Painting, stripping floors, preventive maintenance on heating, ventilation, and air conditioning (HVAC) units, and construction operations can result in IAQ complaints. Depending on your communication system, the target population can be informed through the use of E-mail, computer bulletin board, or memo (internal mail or hand delivered). Whatever the medium, the same method of communication should be used consistently.

It is also important to develop a rapport with the local occupational safety and health officials having jurisdiction in IAQ matters. This should be fostered by continuing communication as well.

Education of the facility staff employees is also very important. As the EPA guidance manual points out, the facility staff employee is in the best position to notice if equipment is not functioning normally or if there is an abnormal event occurring that could affect IAQ. This education process should include training on how to respond verbally to a complaint. Perception plays an important part in identifying an IAQ problem. The smell of sewer gas might not qualify as a problem to a plumber but could cause serious problems in a closed office setting. *Every IAQ complaint must be taken seriously by all individuals involved.*

If a building occupant registers a complaint with a housekeeper, a building-maintenance worker, or the director of safety, and has any indication that she/he is not taken seriously, then all subsequent communications could be of an adversarial nature. Each institution should publish an internal IAQ guideline document that establishes procedures for facilities engineers, maintenance personnel, and the environmental health and safety group to follow in the course of IAQ investigation. The document need not cover all facets of IAQ but should look at the big picture and define common terms, set forth individual responsibilities, and provide background information on common symptoms, signs, and causes of IAQ problems. Administrative support is another very important part of a successful IAQ management program.

Although each IAQ investigation is unique and must be investigated as such, generally, investigations can be divided into three separate phases.

The *first phase* involves a site visit in response to an initial complaint. Information on the potential sources and pathways of pollutants, the status of the HVAC system, and the operations being performed in the area (and in adjacent areas) is collected and evaluated. Also, a simple visual inspection of the complaint area and the surrounding area is performed. Interviews with the occupants are conducted to identify the potential problem areas. Most IAQ complaints are solved with the information collected during the first phase of review and the initial visual inspection. A written record should be maintained.

If the complaint is not resolved to the satisfaction of the concerned parties, and additional investigation is deemed necessary, the evalution process reaches a second phase. If the source of the complaint is not identified or appears to be of a chronic nature, a multidiscipline team should be assembled to continue the investigation. The team should consist of maintenance personnel from the area, a mechanical-engineering component, and representatives from health and safety, the employee health element, and administrative officials having jurisdiction over the area in question.

In the *second phase,* the information-gathering process should build on the information gained in the first phase. Additional interviews and questionnaires can be used.

If symptoms of a nonspecific nature and typical of sick-building syndrome are identified, testing can be conducted for commonly monitored indicators of IAQ. The following basic parameters can be evaluated:

Temperature:	occupant comfort: indicator of effective HVAC operation
Relative humidity:	occupant comfort, organism growth: If the relative humidity is between 30% and 60%, the growth of allergenic or pathogenic organisms tends to be minimized.
Carbon dioxide:	indicator of amount of outside air being supplied to an area: 1,000 parts per million (ppm) is a much-quoted level when evaluating IAQ; personal experience has indicated that complaints tend to disappear when levels are below 500 to 600 ppm.
Particulates:	housekeeping, effectiveness of status of filters for HVAC system: comparisons between noncomplaint areas tend to be most effective when looking at this parameter.
Carbon monoxide:	indicator of sources of combustion

This sampling can be conducted by in-house personnel or with the use of consultants. For a large institution with many indoor air-quality complaints, it is probably more cost effective to develop an in-house sampling program. For a small institution with few complaints it is probably more cost effective to use consultants.

With recent advances in the development of air-monitoring instruments, there are several instruments in widespread use, on the market in about the same price range, that can perform the basic sampling functions described above. Costs for such an instrument could be shared by several institutions in the same geographic area to make the $8,000 to $12,000 price more manageable.

During this phase of the investigation, the ventilation rates, air-flow patterns, and complete pollutant pathways should be evaluated by the mechanical-engineering component (or equivalent) of the IAQ investigation team. This evaluation can range from a full analysis of the HVAC system to simply checking air-flow patterns with smoke tubes. Housekeeping practices also should be evaluated during this phase.

If the source of the complaint is not found and corrected during the second phase, then the investigation reaches a *third phase*. Investigation at this stage should be symptom driven (legionnella, Pontiac fever, hypersensitivity pneumonitis, etc.) and should include investigation of likely causes. Symptoms from building-related illnesses would fall in this area. This could include: microbial testing, sampling for volatile organic compounds, and other pollutants as indicated by the symptoms. As always, testing based on symptoms should be performed based on information gathered from evaluations of the medical status of the building occupants conducted by occupational health physicians or medical clinicians with specialized knowledge in the field of indoor air quality.

Investigators and building occupants need to be clear about what will constitute a successful resolution to an indoor air-quality problem.

Whatever approach is used when addressing IAQ complaints and problems, it is clear that IAQ is another issue that must be faced by institution health and safety departments.

4.26   BIOSAFETY

This section on biosafety is designed to present an overview of the management of biological stressors (biological hazards) that can be encountered in today's workplace environment. Topics covered include descriptions of general categories of biological hazards and generally accepted guidelines for working safely with these various agents.

Virtually everyone is potentially exposed to biologic agents on a daily basis. Food, water, air, humans, insects, animals (zoonoses), and inanimate objects (sharp instruments) are a few common vectors capable of spreading pathogenic (disease-causing) organisms. In most cases, the human immune system is able to prevent the exposure from causing disease.

Biological hazards, or biohazards, consist of pathogenic microorganisms and like substances that could pose a risk to the health and physical well-being of humans, animals, or other biological organisms. In the past, this definition was limited to infectious pathogenic microorganisms responsible for common communicable diseases. The realm of biohazardous agents can now be expanded to include the following agents with the potential for causing disease: oncogenic viruses, recombinant DNA molecules, animals and plants and their by-products, and microorganisms of a rare or exotic nature (fungi, yeasts, algae).

The presence of these biohazards in the environment is inevitable, but certain segments of the population who are occupationally exposed are placed at higher risk of developing disease. This elevated risk is due to increased contact with biohazards, the relatively high quantities of organisms that may be involved during an exposure, or a compromised immune system.

Individuals who are at a higher risk from exposure to biohazards include health-care workers, patients, veterinarians, animal handlers, farmers, dairymen, sanitation workers, emergency-response personnel, laboratory personnel involved in biotechnological research, and individuals with compromised immune systems.

The Centers for Disease Control (CDC) and the National Institutes of Health (NIH) have developed guidelines for different levels of protection when working with biohazards. These levels of protection are divided into Biosafety Levels 1, 2, 3, and 4. Biosafety Level-4 practices and engineering controls offer the highest level of protection.

4.27   BIOSAFETY: CATEGORIES OF HAZARDS

The major categories of biohazards identified by the CDC-NIH publication, *Biosafety in Microbiological and Biomedical Laboratories*, are parasitic agents, fungal agents, bacterial agents, rickettsial agents, and viral agents. Other agents considered biohazards include recombinant DNA products, allergens, potentially infected clinical specimens, and bacteria or plant toxins. Some of these can be broken down into further classifications and specific organisms.

Parasitic Agents

A parasite is an organism that lives on or in a second organism (or host), feeding and multiplying at the host's expense while not contributing to the welfare of the host. Nematode, protozoal, trematode, and cestode parasites have either caused infections in exposed high-risk populations of workers or could do so very easily. Hookworms, *Giardia spp.*, *Cryptosporidia spp.*, and *Schistosoma spp.* are some of the more commonly known parasites of concern when dealing with human exposure. Biosafety Level-2 methods of control are usually warranted when working with the infective stages of the above-listed parasites or similar organisms.

Fungi

Fungi are a ubiquitous, diverse group of organisms. Most are nonpathogenic and are either used in their natural form or to produce other useful products: edible mushrooms; the antibiotic, penicillin; yeast for making bread and beer; and to produce Camembert and Roquefort cheeses. Fungi are made up of eucaryotic cells (complex cells with true nuclei) similar to those found in higher plants and animals. Fungi are like animals in that they are heterotrophic organisms which must consume organic matter to function. Fungi can consume dead organic matter or absorb tissue from living organisms. They absorb their nutrients in the form of a watery solution.

Some fungi can cause mild disorders, such as actinomyces and histoplasmosis. *Aspergillus spp.* and *Cryptococcus spp.* have been identified as potential sources of indoor air-quality problems. Some species of fungi are capable of producing mycotoxins, toxic metabolic by-products. These toxic by-products can occur in a variety of plant foods and in products derived from animals that have eaten contaminated feeds. *Aspergillus flavus* can release the mycotoxin known as aflatoxin, which is the most potent liver carcinogen known to man. Aflatoxin is a frequent contaminant of nuts, grain, and other crops and has been found in food intended for human consumption. Although many fungi can be manipulated using Biosafety Level-2 methods, fungal agents capable of toxin production require Biosafety Level-3 precautions.

Bacteria

Bacteria are small, primitive, one-celled organisms called procaryotes (cells whose genetic material is not enclosed by a nuclear membrane). Although very small (about 0.1 to 0.5 micrometers [μm]), most bacteria have distinctive cell shapes that affect their behavior and persistence. Some bacteria are beneficial (synthesis of vitamins for the body, production of oxygen by photosynthesis), while others act as pathogens to humans or animals. Bacteria have been the source of food- and waterborne diseases (*Staphylococcus aureus, E. coli, C. botulinum, V. cholerae, Salmonella spp.*), and have been identified as a source of indoor environmental problems (*Legionella pneumophilia*). Biosafety Level-2 and -3 methods of protection are sufficient for most moderate-risk bacterial work in a laboratory work setting. Two of the most commonly reported laboratory-acquired bacterial infections, brucellosis and tuberculosis, are considered moderate risk.

Rickettsia

Rickettsial agents are obligate intracellular parasites. This means that they need living host cells in order to multiply. This host-dependence causes them to be generally less hazardous than pathogens with less-stringent survival requirements. However, moderate-risk agents do exist; included in this classification are the rickettsia that cause such diseases as Q fever, Rocky Mountain spotted fever, and typhus. Chlamydias act in a manner similar to rickettsia, causing infections such as psittacosis, a zoonoses transmitted by birds and causing flu-like symptoms in humans.

Viruses

Viruses are small (20 to 300 nanometers), simple genetic structures unable to change or replace their parts. Like the rickettsias, viruses are obligate intracellular parasites needing living host cells in order to multiply and transfer the viral genetic material to other cells. Viruses have been identified that can infect animals, plants, algae, fungi, protozoa, and bacteria. Viruses are classified on the basis of the hosts they infect. Because the nature of viral growth is tied to host cell functions, viruses are difficult to attack specifically using medical therapy without causing some harm to the host cells as well. This results in the relative risks of viruses varying more than any other biohazard category.

Diseases caused by viral agents include influenza, measles, mumps, and yellow fever. Viral agents are also capable of causing human immunodeficiency virus (HIV), hepatitis, polio, and rabies. Although Biosafety Level-2 methods of protection are appropriate for these agents, there are some scenarios where Biosafety Level-3 would be appropriate. The OSHA Bloodborne Pathogen Standard should be followed whenever individuals are working with blood-borne pathogens (HIV, hepatitis B virus).

The *Catalogue of Arboviruses and Certain Other Viruses of Vertebrates* and associated supplements have listed recommended Biosafety Levels for known arboviruses (arthropod-borne viruses) reported to have caused laboratory-associated infections. Most should be handled using at least Biosafety Level-2 precautions and engineering controls. Some examples include: Vesicular stomatitis, Colorado tick fever, and Dengue.

Some viruses that infect normal mammalian cells cause a transformation in the cell that leads to profuse growth. These large masses of cells are called tumors. Tumor formation is referred to as an oncogenic process. Although a direct relationship between viruses and cancer has not yet been positively identified, it is probable that one exists. Oncogenic viruses and resulting tumors are considered moderate risk.

Biosafety Level-4 practices and engineering controls should be used when working with agents such as Congo-Crimean Hemorrhagic fever, Lassa virus, Marburg, and various tick-borne encephalitis virus complexes (Kyasanur Forest disease and Omsk hemmoragic fever). These are commonly associated with a high mortality rate. Biosafety Level-4 practices and engineering controls should also be used with agents that are thought to be similar in nature to these and other similar identified agents. Surviving many of these deadly diseases, which have no known cures, is due only to the ability of the person's immune system to rally while receiving life-supportive therapy.

4.28   BIOSAFETY: RECOMBINANT DNA

A highly controversial research area involves recombinant DNA molecules. This involves the insertion of genetic DNA from one organism to another genetically distinct organism. E. coli is commonly used as the host. The E. coli organism is disabled in some manner to preclude its survival outside of very restrictive growth conditions. Presumably, any release to the environment would result in immediate destruction of the organism. Although this is one of the most common types of recombinant transfer, there are many possibilities. Initially, due to the unknown nature of this research and the overwhelming public concern, NIH developed recommendations for safe handling. These are published in the document "Guidelines for Research Involving Recombinant DNA Molecules." Although these guidelines were originally intended to apply to NIH-funded research, most research institutions and commercial laboratories have voluntarily complied with them. Subsequent research has confirmed that the levels of risk for many recombinant DNA molecules are not as high as originally thought. As a result, the NIH continues to lower the relative risk for many recombinant DNA projects.

New proposed recombinant DNA guidelines have classified biohazardous agents by risk group (see Table 4–1). The relationship of risk groups to biosafety levels, general practices, and equipment is seen in Table 4–2.

Table 4–1. Proposed Recombinant DNA Guidelines

Risk Group 1	No or very low individual and community risk: An agent that is unlikely to cause human disease. Well-characterized agents not known to cause disease in healthy adult humans and of minimal potential hazard to laboratory personnel and the environment.
Risk Group 2	Moderate individual risk, low community risk: Agents that can cause human disease but are unlikely to be a serious hazard to workers, the community, or the environment. Laboratory exposure may cause serious infection but effective treatment and preventative measures are available and the risk of spread of infection is limited.
Risk Group 3	High individual risk, low community risk: Agents that usually cause serious human disease but do not ordinarily spread from one infected individual to another. Effective treatment or preventive measures are available.
Risk Group 4	High individual and high community risk: Agents that can cause serious human disease and can be readily transmitted from one individual to another, directly or indirectly. Effective treatment and preventive measures are usually not available.

4.29 BIOSAFETY: ALLERGENS

Allergens can cause a wide range of diseases in sensitized individuals. Exposure to animal and plant products can result in a variety of adverse clinical responses from those sensitized to allergens. Although avoidance is the best method of control, good housekeeping, ventilation, and as a last resort, respiratory protection can be used to minimize exposure and subsequent clinical manifestations of disease. For individuals who are sensitized to allergens, treatment by medical clinicians specializing in allergies is recommended.

4.30 BIOSAFETY: RISK CLASSIFICATION

The assessment of risk for a particular etiological (disease-causing) agent is subjective and depends in large part on the activities involved during the potential exposure to the biohazardous agent. However, known biohazardous agents have been placed into one of four risk classifications based on their general level of risk to occupationally exposed persons. These four categories of risk were developed by the CDC in the document "Classification of Etiologic Agents on the Basis of Hazard."

The risk-assessment classifications have resulted in the publication of a set

Table 4–2.   Risk Group Comparisons

Risk Group	Biosafety Level 1	Examples of Laboratories	Laboratory Practices	Safety Equipment
1	1	Basic teaching	Laboratory practices	None, open bench work
2	2	Primary health services, primary-level hospital, diagnostic, teaching, and public health	Good micro-biological practices	Open bench plus *BSC for potential aerosol-producing operations
3	3	Special diagnostic	Level 2 plus special clothing, controlled access, directional airflow	BSC and/or other primary containment for all activities
4	4	Dangerous patho-gens units	Level 3 plus airlock-entry shower exit, special-waste disposal	Class-III BSC or positive-pressure suits, double-ended autoclave, filtered air

*BSC: Biosafety Cabinet

of guidelines, *Biosafety in Microbiological and Biomedical Laboratories*, published by the CDC and NIH, which should be used when working with any identified or suspected biohazardous agents. The first set of guidelines was published in 1984, with a third edition published in May 1993. This set of guidelines has become the standard reference guide for determining levels of risk and procedures for working with recognized biohazards.

4.31   BIOSAFETY: GENERAL PRINCIPLES

Biohazards should be controlled using administrative and engineering controls with respiratory protection used as a last resort. Primary controls of biohazards should use different combinations of laboratory practices and techniques and primary barriers (a barrier between the worker and the biological agent). Commonly used primary barriers would include biological safety cabinets, other enclosed containers, and personal protective equipment (PPE). Secondary barriers (special facility design to provide a protective barrier for both individuals working in the laboratory and those outside the work area) should be used as needed to supplement primary barriers.

The Occupational Safety and Health Administration (OSHA) Bloodborne Pathogen Standard (29 CFR 1910.1030) has addressed protective measures for minimizing an individual's potential exposure to blood-borne pathogens (human immunodeficiency virus [HIV] and hepatitis). In December 1991 OSHA published its final rule on occupational exposure to blood-borne pathogens—those microorganisms that may be present in human blood and can cause disease. The primary focus of the regulation is minimizing workplace transmission of both the hepatitis B virus (HBV) and human immunodeficiency virus (HIV). The Standard took effect in stages throughout 1992, and was upheld by the U.S. Court of Appeals for the Seventh Circuit.

The rule applies to all "occupational exposure"—which is defined as *reasonably anticipated* skin, eye, mucous membrane, or parenteral contact with blood or "other potentially infectious materials" that may result from the performance of an employee's duties. Such "duties" can include participation in first-aid teams, rescue squads, or cleanup crews through which employees are expected to provide treatment or otherwise be exposed to blood or other covered materials. Thus, even where a company does not employ in-house doctors or nurses, coverage may be triggered where an employee's duties can lead to occupational exposure.

Each covered employer facility with at least *one* employee with "occupational exposure" must establish a written exposure control plan. Thus, this documentation could be required in the case of a location where the company had organized a first-aid team to provide immediate assistance to employees injured at the work site.

An important provision of the standard is the requirement that the employer offer to all employees occupationally exposed to blood or other bodily fluids free hepatitis B vaccine and vaccination series, and provide post-exposure evaluation and follow-up to all employees who have had an exposure incident. After significant debate, OSHA decided that the vaccination would not be mandatory, primarily out of respect for individual beliefs and rights to privacy. OSHA also rejected its earlier view that vaccination should be based on *average* monthly exposure—ultimately deciding that vaccinations must be offered to *all* individuals with reasonably anticipated occupational exposure.

The Bloodborne Pathogens Standard requires communication of hazards to employees through labels and training. In general, a fluorescent orange or orange-red "BIOHAZARD" label (or other appropriate forms of warning, such as a red bag) must be provided on containers of regulated waste; on refrigerators or freezers that are used to store blood or other potentially infectious materials; and on other containers used to store, dispose of, transport, or ship either blood or other potentially infectious materials.

The employer must provide, to all employees with occupational exposure, free training regarding the hazards associated with blood and other potentially infectious materials and the protective measures to be taken to minimize the risk of occupational exposure. According to OSHA, training will ensure that employees understand hazards associated with blood-borne pathogens, the modes of transmission, the employer's exposure-control plan, and the use of engineering controls, work practices, personal protective clothing, emergencies involving exposure to blood and other potentially infectious materials, and the reasons that they should participate in hepatitis B vaccination and post-exposure evaluation and follow-up.

These protective measures include the use of training and education, good work practices, and personal protective equipment (gloves, masks, lab coats, etc.) to minimize or eliminate infection due to blood-borne pathogens. These protective measures are known collectively as Universal Precautions.

4.32 BIOSAFETY: ADMINISTRATIVE CONTROLS

Education, training, and administration procedures are the most important elements of a successful biosafety program. These essential components apply to all personnel at risk from exposure to a biohazardous agent: from sewage-treatment workers to health-care workers. Although the majority of recommended procedures and practices have been devised for lab personnel, the principles of disinfection, decontamination, and handling have wide applicability and can be used whenever a biohazard is present.

Minimum recommended work practices include:

- No eating, drinking, smoking, or gum chewing is allowed in the area designated for biohazards.
- No mouth pipetting. Only mechanical pipetting should be permitted in areas where there are biohazards.
- No unauthorized persons should be allowed to enter.
- Adequate precautions should be taken to eliminate aerosol formation from processes such as centrifugation, lyophilization, sonication, and grinding. Where possible, these procedures should be performed inside a biosafety cabinet.
- Personal protective equipment should be used whenever appropriate: lab coats, gloves, respiratory protection.
- Prior to disposal, all waste and discarded by-products must be carefully and thoroughly disinfected by autoclaving, incineration, or chemical treatment.
- Animal handling must be done with the same degree of caution as handling the microorganisms directly. In addition, the facility and animal by-products must be disinfected with the same diligence given the main work areas.

- All employees should be under ongoing medical surveillance to identify those who might be at elevated risk.
- Good housekeeping practices are critical for maintaining safe working conditions.
- A Biohazards Safety Committee should be established to promulgate and enforce regulations specific for the hazard.
- The universal biohazard symbol is required by OSHA to be prominently displayed on all biohazardous storage and work areas. It should be displayed on access doors, refrigerators, incubators, animal facilities, equipment, and waste containers.

4.33 BIOSAFETY: ENGINEERING CONTROLS

Effective engineering controls involve preventing biohazardous agents from contaminating the environment or infecting the worker.

One primary engineering-control method involves the separation of biohazard areas from nonbiohazard areas. This separation protects the worker, as well as the research activity, from contamination and exposure. This can be done with various combinations of physical barriers, ventilation, and environmentally enclosed personnel protective suits.

Use of air-pressure gradients can be used to maintain negative air pressures at work areas in relation to surrounding nonwork areas. This is essential for effective engineering control of highly infectious biohazards by preventing their airborne spread from designated work areas. An experienced engineer or industrial hygienist should be consulted when designing such a facility.

It is necessary to filter-exhaust potentially contaminated air with high-efficiency-particulate-air (HEPA) filters. A HEPA filter is minimally designed to remove particulates (0.3 μm in diameter) with a 99.97% efficiency. Equivalent methods of cleaning contaminated air can be used if they are as effective.

To facilitate cleaning and decontamination, all walls, floors, and work surfaces must be smooth and easily cleanable with a minimum of cracks and corners. These design considerations should be applied to both employee work areas and animal quarters.

Decontamination can be accomplished with the use of chemical disinfectants (alcohols, chlorine compounds, formaldehyde, glutaraldehyde, quaternary ammonia compounds), heat sterilization (autoclaves), and sterilizing gases (ethylene oxide, hydrogen peroxide, paraformaldehyde).

Radiation in the form of ultraviolet light can be used for destruction of airborne biohazards (viruses, mycobacteria, bacteria, and fungi) and decontaminating some work surfaces. However, personnel should take care to prevent exposure due to the potential hazards of exposure to ultraviolet radiation.

Design considerations are also essential to prevent the discharge of untreated liquid waste effluent directly into the sanitary sewer or to prevent the accidental flooding of other portions of the facility.

Ventilated biological safety cabinets and ventilated animal cages should be used for all moderate- and high-risk agents and for agents with special hazards. There are three classes for safety cabinets.

Of the three types of biological safety cabinets (BSC), a Class-I BSC provides the minimum level of protection for working with a biological agent. The Class-I BSC is used for low- and moderate-risk agents. The exhaust air either travels through a HEPA filter or is exhausted outside the work area. One disadvantage of using a Class-I BSC is that research materials are subject to contamination because the incoming supply air is not filtered.

Class-II BSCs use HEPA-filtered supply air for product protection and HEPA-filtered exhaust air to protect the environment. The Class-II BSC has replaced the Class-I BSC in many instances. The Class-II BSC is classified according to design parameters, function, and exhaust systems into two different types: IIA and IIB.

The Class-IIA BSC is a stand-alone unit that either recirculates air or exhausts to the outside through a "thimble" connection that enables the cabinet to function even on a failure of an external blower. The Class-IIA BSC should not be used with toxic chemicals or radionuclides.

The Class-IIB is "hard-ducted" to the exhaust system and relies on an external blower to maintain a negative pressure. The Class-IIB BSC can be used with toxic chemicals and radionuclides.

Class-III BSCs are gas tight and provide a total physical barrier, isolating the agent from the surrounding environment. The Class-III BSC receives HEPA filtered supply air and is exhausted through at least two HEPA-filters connected in series. The Class-III BSC offers the highest degree of protection for the worker, the environment, and the research agent.

Biosafety cabinets should be certified to perform according to design specifications on initial installation, any time the BSC is moved, and at least annually thereafter. Certification should be performed by an individual familiar with manufacturers' procedures for certification of applicable biosafety cabinets.

4.34 BIOSAFETY: LEVELS OF PROTECTION

The different combinations of administrative and engineering controls for protection from biohazards are known as Biosafety Levels 1, 2, 3, and 4. Biosafety Level-4 administrative and engineering controls offer the highest level of protection.

Biosafety Level 1 (BSL 1)

Biosafety Level-1 practices and engineering controls are used for agents with minimal or no known hazard. These agents can be handled safely without special apparatus or equipment, using standard microbiological techniques generally acceptable for nonpathogenic materials. Generally, primary containment is provided by adhering to standard laboratory practices during open bench operations. Several basic rules of biological safety (biosafety) should be followed whenever the potential for biological contamination exists:

1. Do not mouth pipette.
2. Wash hands whenever there is a potential for contamination.
3. Use appropriate protective personnel equipment.
4. Handle infectious fluids in such a manner to avoid spills and production of aerosols.
5. Minimize the use of needles and other sharp instruments.
6. Decontaminate work surfaces whenever there is a potential for contamination.
7. Do not eat, drink, smoke, or store food in the work area.
8. Be sensible.

This classification includes all bacterial, fungal, viral, rickettsial, chlamydia, and parasitic agents not included in higher classes. However, some of these agents could be considered opportunistic pathogens and could cause disease in the young, old, and immunosuppressed populations.

Biosafety Level-1 methods of protection are usually appropriate for undergraduate and secondary educational training and teaching laboratories.

Biosafety Level 2 (BSL 2)

Biosafety Level-2 practices and engineering controls are used for agents generally considered moderate risk for potential hazard. Normally, if the potential for producing aerosols with these agents is low, work can be done on the open bench. If the potential for producing aerosols is high, then primary barriers should be used to minimize the risk of infection. Agents in this category may produce disease of varying degrees of severity, but can usually be adequately and safely contained by good laboratory techniques.

In addition to those procedures used during Biosafety Level-1 work, laboratory personnel should be supervised by capable scientists and be proficient in their specific work procedures. Access to the laboratory or work area should be limited during work procedures. Primary barriers should be used when performing operations with infectious agents that could cause aerosols.

Many infectious agents are in this category: hepatitis B virus, measles, mumps, and other viruses.

Biosafety Level-2 methods are appropriate for clinical, diagnostic, teaching, and other similar facilities working with moderate risk agents.

Biosafety Level 3 (BSL 3)

Biosafety Level-3 practices and engineering controls are used for agents where aerosol exposure resulting in infection could cause serious or lethal results. Organisms handled under Biosafety Level 3 require special handling and containment. The personnel should have special training in handling moderate-risk biohazards.

In addition to standard microbiological practices, regular work practices should also include: controlled access to the work area, daily decontamination of work surfaces, decontamination of waste before disposal, hand washing after potential exposures, immediate spill cleanup and decontamination.

Negative air pressure in the work area should be maintained in respect to surrounding areas. Local containment in the form of a biological safety cabinet (Class I or Class II) can be used as a primary barrier. The work area, room, or laboratory, also must be under pressure negative with respect to the remainder of the facility. Air must be decontaminated prior to either discharge to the environment or exhausted through HEPA filters.

Animal experiments (including cage sterilization, refuse handling, and disposal of animals and animal parts) must be conducted with equivalent levels of precaution, and all waste must be decontaminated prior to disposal.

Protective clothing and gloves must be worn and decontaminated. Respiratory protection should be used as needed.

Personnel must be immunized or, if possible, placed into an appropriate medical-surveillance program.

Biosafety Level 4 (BSL 4)

These agents require the most stringent conditions for containment due to their extremely dangerous and exotic nature. These organisms can be transmitted by the aerosol route and there is no available medical intervention (cure or vaccine). These organisms are highly infectious and/or capable of causing debilitating or life-threatening disease. The minimum containment conditions and work practices for these agents are the same as for those treated with Biosafety Level-3 methods of protection plus the following:

Work areas should either be in a separate building, or environmentally isolated from surrounding areas by the use of ventilation and enclosed waste-management systems. Workers should change clothing when entering and shower out of the work area.

All work can be handled only under conditions of total containment. This can be done by either containing the biohazard in closed-environment glove boxes (Class III BSC) or by using a Class-I or Class-II BSC with the worker enclosed in a one-piece environmentally isolated positive-pressure suit. The suit must include a life-support system that totally isolates the individual from the contaminated environment.

4.35 FIRE SAFETY

Fire prevention and protection is a very integral component of an institutional health-and-safety program. The primary goal of a fire-prevention and -protection program is to protect life. Secondary to this goal is protecting physical property and maintaining continuity of operation.

A systematic approach to fire safety is necessary for an effective fire-protection program. There are six major elements that should be incorporated:

- prevention of ignition
- design to slow early fire growth
- detection and alarm
- suppression
- confinement of fire
- evacuation of occupants

The first opportunity to achieve fire safety in a building is through fire prevention. Major factors in fire prevention include heat sources, forms and types of ignitable materials, circumstances bringing heat and ignitable material together (i.e., misuse of heat source or material, equipment failure), and practices affecting prevention success (i.e., housekeeping, control of fuel type, education).

When new buildings or additions are planned, the project management team should emphasize to the architect that the design principles listed above be considered.

Fire-detection provisions are necessary so that automatic or manual fire suppression will be activated and occupants will have time to evacuate.

Automatic sprinklers are the most important engineering-control system for automatic control of fires in buildings.

Barriers, such as walls, partitions, and floors, delay or prevent fire from propagating from one space to another. They also help to prevent the structural collapse of the building during a fire.

The evacuation of occupants may involve one or a combination of the following three alternatives:

1. evacuation of occupants
2. defending the occupants in place
3. providing an effective area of refuge

Evacuation involves the availability of acceptable means of egress and the effective alerting of the occupants in sufficient time to allow egress before segments of the path of egress is unavailable.

The second design alternative is to defend the individual in place. This may be appropriate for occupancies such as hospitals, nursing homes, prisons, and other institutions.

The third alternative is designating an area of refuge. This involves occupant movement through the building to specifically designed refuge spaces. This is used frequently to help with disabled-person-accommodation requirements.

The minimum frequency of inspections for fire safety is mandated by national and local regulations and guidelines. However, it is prudent to evaluate occupancies according to use, population density, fire-department access, fuel loading, and built-in fire-detection and -protection systems and develop increased inspection frequencies for those occupancies that are at a higher risk.

Communication with the local fire department concerning building information is helpful in the fire department's ability to extinguish a fire with minimal threat to life and safety.

Factory Mutual Engineering and Research (1992) loss statistics indicate that electrical hazards, arson, and smoking caused 69% of insured college and university fire losses in the past ten years. These three common causes accounted for 83% of the gross dollar loss during that same period. NFPA statistics reveal that during the 1970s to 1980s, 88% of major college and university fires involving fatalities occurred in student residential buildings (including fraternities and sororities).

Building and fire codes are used to provide fire-safety requirements for contruction. Typical building codes used throughout the United States are the American Insurance Association (AIA), the International Conference of Building Officials (ICBO), the Building Officials and Code Administrators (BOCA), and the Standard Building Code (SBC). The most widely used fire codes include the National Fire Protection Association (NFPA), the Basic Fire Prevention Code (BFPC), ICBO's Uniform Fire Code (UFC), the Southern Standard Fire Prevention Code, and AIA's Fire Prevention Code.

Local officials having jurisdiction will sometimes interpret national codes to fit their particular jurisdiction. Whenever there appears to be some disagreement with interpretation, verification should be requested in writing. One effective method of eliciting a favorable response is to send your interpretation and support documentation and ask for an answer if there is disagreement with the interpretation. Always send via registered mail and include a deadline for response. Make sure the deadline-response date gives the responding authority ample time to respond if a response is necessary. It is a good idea to have a building profile on hand, specifying what particular set of codes are applicable (year and type). Many inspection officials will inspect a building citing current codes when the building was built under a different set of codes.

It is important to maintain ongoing communications with local fire officials to alert them to potential problem areas and any special fire-fighting procedures that could be needed.

It is a good idea to take the local firefighters on tours of your institution to familiarize them with the physical layout they will be encountering during an emergency situation. At this time it would be a good idea to find out what the capabilities of the local group is in other related emergency areas. Frequently, the fire department will be the first personnel on site in the event of a spill or confined-space accident.

4.36   OSHA RECORD-KEEPING REQUIREMENTS

When Congress passed the Occupational Safety and Health Act, it provided authority for the Secretary of Labor to promulgate rules regulating an employer's conduct in operation of their businesses. Section 6 of the Act gave the Secretary general rule-making authority with specific procedural guidelines, as well guidelines on the content of the rules Occupational Safety and Health Administration (OSHA) could impose. Congress specifically authorized OSHA to include provisions requiring employers to maintain records for various purposes. This section will discuss the statutory basis for OSHA's authority and then identify the kinds of records that must be kept, the retention period, and the rules regarding access to the records by various parties.

Congress gave OSHA authority to regulate employers' actions that could affect the health and safety of employees. Recognizing that government must have information with which to be effective, Congress specifically authorized OSHA to require employees to keep records. Specifically, the statute provides:

> Each employer shall make, keep, and preserve, and make available to the Secretary . . . such records regarding his activities relating to this Act as the Secretary . . . may prescribe. . . .
> The Secretary . . . shall prescribe regulations requiring employers to

maintain accurate records of, and to make periodic reports on, work-related deaths, injuries, and illnesses. . . .

The Secretary . . . shall issue regulations requiring employers to maintain accurate records of employee exposures to potentially toxic materials or harmful physical agents. . . .

A further reason for OSHA to require records is the need to document the history of the employee's work environment for future research and studies. One of the major reasons that Congress enacted the statute in 1970 was the recognition that some kinds of work-related illnesses develop after exposure over long periods of time, and that little information was available in 1970 to define "safe" working conditions on the basis of objective science.

There is a more basic reason for the requirement to keep records, which also explains the tendency of OSHA to continually increase these requirements. That reason is the need for the Agency in its enforcement posture to document the existence of violations of the rules it issues. It is not possible for OSHA inspectors to be present on the job site every day, or even to inspect every workplace each year. It thus falls to the inspector to ascertain the compliance status of the employer through other means. Principal among these is the inspection of the records kept in the ordinary course of business. This is the classic approach of the government enforcers in searching the paper trail for evidence of wrongdoing.

The kinds of records kept generally fall into four classes. *First,* there are the usual kinds of communications between individuals both within the organization and with those outside the organization. These document the steps the responsible manager takes to carry out company policies and activities.

The *second* type of records kept are the written expressions of institutional policies, procedures, and instructions. These document the specific actions that the university, hospital, or school takes to implement both mandated and voluntary programs. In some cases, the statutes or regulations require that a written program be developed; in others, it is simply good management practice to prepare such documents to consolidate the instructions to company personnel.

The *third* kind of records are those documents that describe the specific day-to-day activities of employees in carrying out the mandates of programs and policies. These include daily, weekday, or monthly inspection reports; summary management reports; records of training and discipline; and similar documents.

The *fourth* kind of records, records of activities and programs, are always a double-edged sword. On the one hand, they document the reasonable and prudent actions of company officials and employees in conducting the business lawfully. They also often document the failure of company personnel to perform specific tasks, or of the corporate organization to respond adequately to problems. This tension between the utility of records and the risks inherent in keeping them has created a significant dilemma for many people. Nevertheless, maintenance of good records is always prudent.

The OSHA standards often include requirements for the preparation and maintenance of records. In some cases, records requirements are inferred from the performance-oriented nature of the standard.

The Hazard Communication Standard (HCS) covers virtually all employers in the private sector in the United States with exception of the mining industry. The standard requires employers to address all hazardous chemicals to which employees may be exposed. All employers are required to communicate to their employees information about hazardous substances which are known to be present at the work site. This requirement applies regardless of whether the employer created the exposure. The issue for all employers is whether they "know" that their employees are exposed.

Chemical manufacturers and importers are required to perform hazard determinations on all chemicals they produce or import. If downstream employers, such as wholesalers and distributors, re-label products or in any other way choose not to rely on the manufacturer's determination of hazard, they too must perform the determination.

4.37   HAZARD COMMUNICATION STANDARD: RECORDS

There are four specific kinds of records that must be kept under the HCS. *First,* there is the written program. This must contain sections that address specific subjects listed in the standard (CFR 1910.1200(e)). *Second,* every employer must keep a list of hazardous chemicals to which employees may be exposed. This list must contain the names of the products that are used on the label and material-safety data sheets (MSDS). The user must be able to find the appropriate MSDS from the name on the label. Consequently, it is *not* clear that the OSHA standard requires the inclusion of component chemical names on the list of hazardous chemicals. This is a common misconception.

The *third* kind of record that must be kept is the MSDS. The ostensible purpose of the MSDS is to provide information to employees about the hazardous chemicals to which they are exposed. However, OSHA attributes a secondary purpose to both the list and the MSDS: to document the exposures of employees over time for use as a tool in conducting epidemiological studies. Thus, these records are considered records of exposure which are subject to the record-retention provisions of 1910.20(d)(1)(ii).

Fourth, labels may also be considered records, although they are arguably not exposure records. Certainly companies will want to keep records of what label statements were used and in particular the reasons for making the choices. The standard requires that "appropriate hazard warnings" be used on labels on all containers in the workplace. OSHA recognizes that appropriate labels need not include warnings about every toxic effect of every component in a product (OSHA CPL). Therefore, where labels evolve as manufacturers learn more about their products, both the inherent dangers and the usage characteristics of customers' operations, the "appropriate" warnings will change. It is important for manufacturers to document this evolution, both for OSHA as well as for other legal purposes, and such records will thus become part of the business's ordinary records system.

These are the records that are explicitly mandated by the HCS, and not all of them are subject to any particular retention policy. However, implicit in the requirements of the standard, such as training, hazard communication, and others, is the need to document compliance activities.

Implementation of the written program necessitates the generation of additional records, such as historical records of program reviews, notes on hazard determinations, drafts of labels, and material-safety data sheets and training records. These records are almost always necessary to document that the program is in fact implemented. For example, only in the case of the employer who relies solely on his suppliers for MSDS and labels will there be no hazard determinations. Unless an employer can demonstrate that employees are well trained by other means, records of training will be necessary.

The last kind of record the HCS implicitly requires is the response of the employer to requests for copies of MSDS or other information by medical specialists in an emergency. Under the standard, manufacturers may withhold chemical composition information from labels and MSDS, but must disclose the information when a bona fide request is received from a physician or other health-care professional. These requests must be in writing and must contain specific information, and the responses must be prepared accordingly. Of course, such correspondence inevitably creates a record of the transaction which then must be maintained.

It is not always obvious that records are being developed when managers perform their jobs. The process of requesting material safety data sheets generates additional records that OSHA inspectors wish to see. These documents are evidence of good faith in carrying out an employer's responsibilities under the HCS. In addition, they leave an audit trail for in-house verifications that compliance programs are effective.

Audits generate additional records that demonstrate both compliance and corporate commitment. These documents are not primary sources of compliance information, but rather document the internal feedback systems necessary for management to assure that company policies and programs are effectively implemented (O'Reilly 1994).

In developing records under the HCS, it becomes important to consider what information must be kept. With regard to hazard determinations, chemical manufacturers should document the decision-making process that they follow to prepare MSDS and labels. The record should reflect the specific sources of information considered, the issues related to selection of hazard-warning statements, and considerations of normal use or foreseeable emergency.

The importance of these documents is that they establish both the employer's good-faith efforts to evaluate the hazard as well as the rationale for selecting the particular content and wording of labels and MSDS. In an inspection situation, the compliance officer is at a distinct disadvantage because he does not have reference sources available. Where the procedures and decisions are adequately documented, he is not in a position to question them unless he is willing to spend a significant amount of time researching the issues. Moreover, the judgment of hazard is the employer's, which, of course, can be subject to review by OSHA, but the burden is then on OSHA to demonstrate that the employer's determination is incorrect.

As can be seen from the above, OSHA sometimes is very explicit about what is expected. In the HCS, the definition of the written program elements is detailed in very simple language:

> Employers shall develop, implement, and maintain a written hazard-communication program for their workplaces which at least describes how the criteria specified in paragraphs (f), (g), and (h) of this section . . . will be met. . . .

To satisfy the standard of describing how these requirements are met requires an extensive written program. Simply paraphrasing the language of the standard is not sufficient. OSHA expects to see details such as responsible managers, descriptions, and procedures written out. In the OSHA Compliance Instruction (CPL) CPL 2-2.38C, *Inspection Procedures for the Hazard Communication Standard*, OSHA defined what it expects to see in a written Hazard Communication Program (HCP).

> The CSHO shall determine whether or not the employer has addressed the issues in sufficient detail to ensure that a comprehensive approach to hazard communication has been developed.

In general, the written program should consider the following elements where applicable:

Labels and Other Forms of Warning
- designation of person(s) responsible for ensuring labeling of in-plant containers
- designation of person(s) responsible for ensuring labeling on shipped containers
- description of labeling system(s) used

- description of written alternatives to labeling of in-plant containers, where applicable
- procedures to review and update label information when necessary

Material-Safety Data Sheets (MSDS)

- designation of person(s) responsible for obtaining/maintaining the MSDS
- how such sheets are to be maintained (e.g., in notebooks in the work area(s), via a computer terminal, in a pickup truck at the job site, via telefax) and how employees obtain access to them
- procedure to follow when the MSDS is not received at the time of the first shipment
- For chemical manufacturers or importers, procedures for updating the MSDS when new and significant health information is found

Training

- designation of person(s) responsible for conducting training
- format of the program to be used (audiovisuals, classroom instruction, etc.)
- elements of the training program—compare to the elements required by the HCS (paragraph (h))
- procedures to train new employees at the time of their initial assignment and to train employees when a new hazard is introduced into the workplace
- procedures to train employees of new hazards they may be exposed to when working on or near another employer's worksite (i.e., hazards introduced by other employees)
- guidelines on training programs prepared by the OSHA Office of Training and Education entitled "Voluntary Training Guidelines" (49 Fed. Reg. 30290, July 27, 1984) can be used to provide general information on what constitutes a good training program

Additional Topics to Be Reviewed

- Does a list of the hazardous chemicals exist and if so, is it compiled for each work area or for the entire work site and kept in a central location?
- Are methods the employer will use to inform employees of the hazards of *nonroutine* tasks outlined?
- Are employees informed of the hazards associated with chemicals contained in unlabeled pipes in their work areas?
- Does the plan include the methods the employer will use at multi-employer work sites to inform other employers of any precautionary measures that need to be taken to protect their employees?
- For multi-employer workplaces, are the methods the employer will use to inform the other employer(s) of the labeling system used described?
- Is the written program made available to employees and their designated representatives?

The HCS therefore imposes a comprehensive duty on employers to develop written documents both explicitly and implicitly.

4.38   COMPLIANCE WITH OTHER OSHA STANDARDS

Other OSHA standards are equally broad. The chemical specific standards often include requirements for the development of specific exposure and medical records, administrative memoranda, reports, and certifications. Table 4–3 lists the chemical specific standards presently in effect. The lead standard, for example, requires the following records be kept:

(d)(2)	initial determination
(d)(4)	initial monitoring
(d)(5)	negative initial determination record
(d)(7)	additional monitoring
(e)(1)	infeasible engineering controls
(e)(3)	written compliance program
(e)(5)	measurements of mechanical ventilation
(e)(6)	job-rotation schedules
(f)(1)	written respirator program
(f)(2)	respirator selection
(f)(3)	respirator fit testing
(g)(2)(vi)	notice to laundries
(g)(2)(vii)	laundry-container labels
(j)(2)	blood lead monitoring
(j)(2)(iii)	employee blood lead notifications
(j)(3)	medical examinations
(j)(3)(iii)(B)	second-opinion notifications
(j)(3)(iv)	information to physicians
(j)(3)(v)	written medical opinions
(l)(1)(ii)	training program
(l)(1)(iii)	initial training
(l)(1)(iv)	annual training
(l)(2)(i)	employee requests for information
(l)(2)(ii)	DOL requests for information
(m)(2)(ii)	sign cleaning
(n)(1)	exposure monitoring record keeping
(n)(2)	medical-surveillance record keeping
(n)(3)	medical removal record keeping
(o)(2)(ii)(c)	observation of monitoring

Table 4–3.   Chemical Specific Standards*

Asbestos	1001	Coal Tar Pitch Volatiles	1002
4-Nitrobiphenyl	1003	alpha-Naphthylamine	1004
Methyl chloromethyl ether	1006	3,3'-dichlorobenzidine	1007
bis-Chloromethyl ether	1008	beta-Naphthylamine	1009
Benzidine	1010	4-Aminodiphenyl	1011
Ethyleneimine	1012	beta-Propiolactone	1013
2-Acetylaminofluorene	1014	4-Dimethylaminoazobenzene	1015
N-Nitrosodimethylamine	1016	Vinyl chloride	1017
Inorganic arsenic	1018	Lead	1025
Benzene	1028	Coke oven emissions	1029
Cotton dust	1043	1,2-dibromo-3-chloropropane	1044
Acrylonitrile	1045	Ethylene oxide	1047
Formaldehyde	1048		

*NOTE: This list was derived from several subsections of 29 C.F.R. § 1910 (1995). Please refer to any subsequent supplements to part 1910 for the most up-to-date information.

In summary, there are 29 provisions in the lead standard in which some form of record is implied or required. This is, of course, an enormous burden, but more importantly, it creates a significant problem for assuring compliance. The data for those records must be recorded and then they must be maintained. This is interpreted by OSHA to mean that the information in the records must be updated when it changes.

Some of the OSHA standards that imply significant recordkeeping requirements do so because they require some kind of periodic inspection. For example, the standard on *portable wood ladders* requires that

Ladders shall be inspected frequently and those which have developed defects shall be withdrawn from service. . . ."[1]

This provision does not explicitly require records. However, most employers would find them useful if an inspector found an employee using a ladder that had a defect. The record could show a recent defect, perhaps, that arose after the last inspection.

---

[1] 29 C.F.R. 1910.25(x).

Table 4–4. Current Requirements for Health and Safety Records in the Workplace

	Written Program	Specific Training	Routine Inspections	1910 Section
Portable ladders	N	N	Y	25
Emergency response	Y	Y	N	38
Walking working surfaces	N	N	Y	68
Ventilation systems	N	N	Y	94
Noise	N	Y	Y	95
Ionizing radiation	N	Y	Y	96
Flammable/combustible liquids	N	N	N	106
Hazardous waste	Y	Y	N	120
Personal protective equipment	Y	Y	N	134
Lockout/tagout	Y	Y	Y	147
Medical services and first aid	N	Y	N	151
Fire brigades	Y	Y	Y	156
Fire extinguisher	Y	Y	Y	157
Sprinklers/hoses/standpipes	Y?	Y?	Y	158
Employee alarms	Y	Y	Y	165
Servicing truck tires	N	Y	N	177
Powered industrial trucks	N	Y	Y	178
Cranes	N	N	Y	178,179
Derricks	N	Y	Y	181
Slings	N	N	Y	184
Machine guarding	N	N	Y	217
Mechanical power presses	N	Y	Y	217
Portable power tools	N	N	Y?	241-244
Welding	N	Y	N	252
Electrical systems	N	Y	N	301-399
Hazard communication	Y	Y	Y	1200
Laboratories (nonproduction)	Y	Y	N	1450
Process safety management	Y	Y	Y	119
Bloodborne pathogens	Y	Y	Y	1030
Confined space entry	Y	Y	Y?	146

Key: Y = required, N = not required.

Table 4–5. Anticipated Requirements for Health and Safety Records in the Workplace

	Written Program	Specific Training	Routine Inspections
Ergonomics	Y	Y	N
Motor vehicle occupant safety	Y	Y	Y

Key: Y = required, N = not required.

The defense to such a citation would likely be the defense of "employee misconduct," that the employee failed to perform his job properly. The employer would have to show that he had a policy requiring employees to inspect the ladders "frequently," that the employee had been told of the policy, and that the policy was actively enforced. Contemporaneous records may show that (1) there was a policy, (2) the employee had been trained, (3) inspections had been conducted regularly by others, and (4) supervisors had enforced the requirement. This last element would likely be shown through evidence that employees had been disciplined for failure to perform the required acts.

This rule could be interpreted to require that at least four types of records would have to be kept to demonstrate employer compliance with the inspection requirement. Of course, not all employers will have such records, and in fact, many do not keep them. Nevertheless, because the burden is on the employer of demonstrating compliance with the standard when the prima facie case is made in the Agency's complaint, records of such kinds of activities are always desirable.

Table 4–4 is a listing, by no means complete, of other OSHA standards that have either written program, training, or inspection requirements. Nearly every standard that is proposed by OSHA today has some provision for additional records. Table 4–5 lists those standards proposed by OSHA that have significant recordkeeping provisions included. We can anticipate that all future standards will be equally designed to provide an adequate record for inspectors to use in evaluating compliance.

4.39 RETENTION OF AND ACCESS TO RECORDS

OSHA does not have a generic record-keeping standard. There is a single provision that addresses retention periods and authorized access to medical records. The retention provisions require employers to maintain medical and exposure records for 30 years, and provide for the transfer of records to the National Institute for Occupational Safety and Health (NIOSH) in the event the employee intends to dispose of them or ceases to do business. Medical records are defined to exempt health-insurance claims records, first-aid records of one-time treatment and subsequent observation of cases not involving medical treatment, and records of employees employed for less than one year who are given the records on termination of employment. The exemption for first-aid records applies essentially to those cases which would otherwise not be recordable on the OSHA Form 200.

Paragraph (e) of this section establishes the right of employees and their designated representatives to obtain access to and copies of medical and exposure records. Employers may not charge for initial copies of records, and must provide access within 15 working days of a request. Employee medical records are subject to a provision that can limit access by the employee if, in the opinion of an employer's physician, there is information regarding a terminal illness or psychiatric condition that could be detrimental to the employee's health. In such cases, the employer may provide the information to another physician of the employee's choosing after denying the employee access in writing to the detrimental information. Confidential information identifying persons who have provided information about the employee may be excised from the record provided to the employee.

A separate regulation authorizes OSHA access to the records as authorized by the statute (CFR 1910.20). These regulations define how employees must be given access to both medical records and measurements of chemical exposures. The rules, issued to satisfy privacy concerns, limit the types of requests that can be made, the agency personnel who may be granted authority to access the records, and limits the uses to which the records are put.

As OSHA compliance becomes more complex, increasing numbers of standards require the completion and maintenance of records. The challenge for employers today is to assure that the records are being kept and that they are accurate, and it is by no means an easy task.

REFERENCES

ACGIH. 1985. Threshold Limit Values and Biological Exposure Indices for 1985–86. Cincinnati: American Conference of Governmental Industrial Hygienists.

American Industrial Hygiene Association Biohazards Committee. 1993. Biohazards Reference Manual. Fairfax, VA: AIHA

ASHRAE. 1981. Thermal Environmental Conditions for Human Occupancy. Atlanta: American Society of Heating, Refrigeration and Air-Conditioning Engineers, Inc. ASHRAE 55-1981.

ASHRAE. 1989. Ventilation for Acceptable Indoor Air Quality. Atlanta: American Society of Heating, Refrigeration and Air-Conditioning Engineers, Inc. ASHRAE 62-1989.

Buckley v *Kruger-Benson-Ziemer* (1987).

(CFR (in text) means Code of Federal Regulations)

29 Code of Federal Regulations § 1910.20.

29 Code of Federal Regulations § 1910.1001.

29 Code of Federal Regulations § 1910.1200.

Engineering Foundation Conference. 1972. Improving Indoor Air Quality. Berwick Academy, South Berwick, Maine.

EPA. 1991. Building Air Quality, A Guide for Building Owners and Facility Managers. EPA/400/1-91/033, DHHS (NIOSH) Publication No. 91-114, Washington, DC.

Fleming, Richardson, Tulis, Vesley. 1995. *Laboratory Safety Principles and Practices.* Washington, DC: ASM Press.

General Carbon Co. v *OSHA,* 860 F.2d 479 (D.C. Cir. 1988).

Her Majesty's Stationary Office. 1954. Mortality and Morbidity During the London Fog of December 1952. London.

Moschandreas, D.J. 1981. Exposure to Pollutants and Daily Time Budgets of People, Symposium on Health Aspects of Indoor Air Pollution, Bulletin NY Academy of Medicine, New York.

Ontario Ministry of Labour. 1988. Report of the Inter-Ministerial Committee on Indoor Air Quality, Toronto.

O'Reilly, James. 1994. Environmental Audit Privilege. *Seton Hall Legislative Journal* 19:119.

OSHA. CPL 2-2.38C, Inspection Procedures for the Hazard Communication Standard at A-20.

OSHA. 1994. Fall Protection Standard. Federal Register 59:40672 (August 9, 1994).

Peterson Brothers Steel Erection Company v. *Robert B. Reich, Secretary of Labor and the Occupational Safety and Health Review Commission,* 26 F.3d 573 (5th Cir., 1994).

Richmond and McKinney. 1974. *Biosafety in Microbiological and Biomedical Laboratories.* U.S. Department of Health, Education, and Welfare, Public Health Service, Washington, DC.

Robinson, J.P. 1977. Changes in Americans' Use of Time: 1965–1975, A Progress Report, Communications Research Center, Cleveland State University, Cleveland.

Salvaggio, J. 1990. The Impact of Allergy and Immunology on Our Expanding Industrial Environment. *Journal of Allergy and Clinical Immunology.*

Schrenk, H., Heimann, J.V., and Clayton, C. 1948. Air Pollution in Donora, PA: Epidemiology of Unusual Smog Episode of October. *Public Health Bulletin.*

U.S. Department of Health, Education, and Welfare, Public Health Service, 1974.
    Classification of Etiologic Agents on the Basis of Hazard, Washington, DC.
Walkinshaw, D.S. 1988. The Sick Building Syndrome. Presentation to Ontario
    Association of School Business Officials' Operations, Maintenance and
    Construction Workshop: Indoor Air Quality in Schools, Kitchener, Ontario.

Chapter **5**

Air-Emission Issues

5.1 OVERVIEW

The Clean Air Act (CAA) (USC, § 7401 *et seq.*) and its huge set of complicated EPA-implementing regulations affect the activities of virtually all educational and health-care organizations, regardless of profit status. Managers of health-care facilities and educational institutions face an array of new requirements with a variety of deadlines during the last half of the 1990s, as a result of the 1990 Clean Air Act amendments. Many facilities will need to obtain or renew Clean Air Act operating permits issued by state regulatory officials under the new federal commands.

For example, facilities may be subject to the CAA's general air-pollution prevention and control provisions if they operate incinerators, generate electricity, use solvents such as those associated with machine or vehicle maintenance, maintain spray-painting operations, or use coal or oil to produce heat. Facilities that operate dry cleaners, steam-generating units, incinerators, grain elevators, or rotogravure printing presses may be subject to the CAA's requirements for specific pollutants or types of sources.

Other activities triggering CAA regulation include:

- maintaining ten or more vehicles (fleet requirements and emission standards for motor vehicles)
- cogenerating steam and electricity and selling electrical output (acid-deposition-control provisions)
- maintaining air conditioners in motor vehicles or any appliance that refrigerates or cools buildings (stratospheric-ozone-protection requirements for compounds such as chlorofluorocarbons (CFCs))

All these topics are addressed in this chapter.

5.2   OPERATING PERMITS

While Clean Air Act permits have been required by federal law for over 20 years, in 1990, Congress directed EPA to overhaul the system and develop a program for regulating more stationary sources of air emissions through a permit mechanism that states would directly administer (USC, § 7661a(b)). Most educational and health-care facilities must obtain permits, and all facility managers should carefully evaluate the applicability of various air regulations (O'Reilly et al. 1995).

Prior to 1990, sources were not required to obtain comprehensive permits for their air emissions, as they are required to do for water discharges. Air permits will now list each source's pollution-control requirements in one document; these were previously scattered among state implementation plans (SIPs), new-source performance standards (NSPS), national emission standards for hazardous air pollutants (NESHAP), and construction permits. The permit program applies to all major stationary sources, regardless of location, and to other sources subject to NSPS, NESHAP, prevention-of-significant-deterioration (PSD) permits, or acid-deposition requirements. EPA regulations establish minimum requirements for state-operating permit programs (USC, § 7661(2)), however, the permitting authority (either a state, local agency, or tribe) may be more stringent.

Individual sources must determine whether they need an air permit. A source is defined as all emissions units within a contiguous area or on adjacent properties, under common control, belonging to a single major industrial grouping. First, a source must identify its air emissions. Does the source emit any "criteria pollutants" or hazardous air pollutants? If so, then it may need a permit.

Criteria pollutants are:

- ozone (which is regulated by restricting emissions of ozone precursors such as volatile organic compounds (VOCs)), *e.g.,* solvents
- NO_x—*e.g.,* from boilers and/or other combustion sources
- SO_2—*e.g.,* from boilers and/or other combustion sources
- CO—*e.g.,* from boilers and/or other combustion sources
- PM10—e.g., from materials handling and incineration

Does the source emit any of the 189 listed hazardous air pollutants (HAPs)? If so, then it may need a permit. Many HAPs are also VOCs and count toward VOC emissions.

Emissions of ozone-depleting substances are not subject to permitting. Chemical inventories, data from existing permits, emissions monitoring, and knowledge of machinery and processes may help hospitals and universities determine what they emit.

Second, a source must calculate the amount of its air emissions. Whether a hospital or university is required to obtain a permit because of emissions of a criteria pollutant depends on how much of the criteria pollutant is emitted or potentially emitted and where the hospital or university is located. In calculating emissions, a source must not forget to include fugitive emissions where appropriate. These emissions are caused by volatilization from open vessels, waste-treatment facilities, spills, and/or shipping containers; leaks from pumps, valves, and/or flanges; or building ventilation systems.

If the source determines that it needs a permit, it must complete a permit application (USC, § 7661b). Sources must submit permit applications within one year of approval by EPA of the state program, or one year from commencing operation, whichever is later (USC, § 7661b). The state has 18 months to issue or deny a permit after an application is filed. Operation without a permit is a violation, but a filed permit application is a shield, provided the application submission is timely submitted and complete. This application shield means that a facility cannot be found in violation of the Act for operating without a permit while the permitting authority is reviewing its application, as long as its operations are consistent with its application. A facility could lose the shield if the permitting authority, in the course of processing the application, requests additional information and the facility does not provide it in a timely fashion.

Permit applications must include:

- identifying information for a source (location, description of processes, emissions)
- applicable air-pollution-control requirements
- a compliance plan and schedule
- a compliance certification
- enhanced monitoring/periodic monitoring
- record-keeping and reporting requirements
- a certification of accuracy

The compliance plan, for sources already in compliance, must include a statement of intent to continue complying; for noncomplying sources, it must include a compliance schedule, with enforceable interim milestones. EPA, or an affected state, has 45 days to veto a proposed permit. Citizens may object to a permit by filing a petition with EPA.

Once a permit is issued, a facility may be protected by a "permit shield" if it is operating in compliance with its approved operating permit. A permit shield means that compliance with the terms of the permit will be deemed compliance with the requirements of the CAA by the permitting authority (USC, § 7661c(f)). A permit shield will apply only if all requirements applicable to the source are included in the permit and the permitting authority determines in writing that other requirements are not applicable to the source and such determination is in the permit. States are authorized to include a permit shield in their operating permits program, but are not obligated to incorporate it. A permit that does not expressly provide for a permit shield is presumed not to include one. The permit must include all applicable requirements, including emissions standards, testing, monitoring, record keeping, and reporting to the state (semiannual for all, prompt for violations) (USC, § 7661c). Additional state requirements may be imposed, but will be designated as not federally enforceable.

All permitted sources must submit an annual statement indicating their compliance status (USC, § 7661b(b)(2)). The statement must be certified by a responsible official of the source and submitted to EPA and the state. For example, a University's Senior Vice President might need to be extensively briefed about the facility emissions before signing the certification statement for the university.

Permits are issued for a fixed term, not to exceed five years. This is a significant change for facilities located in states that did not have renewal requirements and permitted facilities only at start-up and when major modifications took place. Changes in the facility's operations during the five-year life of the permit may trigger permit modification or notification requirements. Any change in facility operations should be viewed as potentially requiring permit modification or reporting until experience is gained under the new rules. Indeed, in drafting permit applications, facility managers should view the permit as a five-year plan and try to include potential expansions or changes within the described permit activity. This, however, is a balancing act. On one hand, there is a distinct advantage to avoiding the need to apply for permit modifications during the permit's five-year life. On the other hand, it is counterproductive to face increased compliance costs for activities that may not occur or may not occur for several years. Thus, careful assessment and planning is warranted.

As this chapter is being prepared for publication, EPA has a pending rule which may simplify permit modification requirements (EPA, August 1994). Thus, facility managers should verify the latest applicable federal, state, and local requirements. Under existing rules, a permit may be modified, but procedures differ depending on whether the modification is: (a) clerical/typographical (does not require public comment); (b) minor (requires public notice, but the source can go ahead with the change at its own risk); or (c) significant.

In general, significant changes in a facility's air emissions require public notice; the source may not make any changes until it receives approval from the permitting authority. Significant changes include all monitoring, record-keeping, and reporting changes. Sources have the flexibility to make certain minor changes in the operation of their facility without reopening their permit. The changes, however, cannot be "modifications" under any of the CAA's criteria pollutant provisions and must not exceed the emissions allowable under the permit. The source must notify EPA and the permitting authority seven days prior to such minor changes. Permits must be reopened to incorporate new requirements if more than three years remain on the permit term or at any time for material mistakes or to assure compliance (USC, § 7661(b)(9)).

A source must make certain to check the state or local requirements that might apply to it. Individual states may have more stringent requirements than EPA's.

5.3   REQUIREMENTS BASED ON LOCATION

One set of CAA restrictions is based on where the facility is located and what pollutants it emits. Basically, U.S. EPA established air-quality standards for six substances. States must create a plan, called a State Implementation Plan (SIP), which describes how the state will maintain or achieve these air-quality criteria. The state plan and accompanying rules will describe practices, restrictions, or controls that various types of sources must follow based on their location within the state.

Air-Quality Standards

Criteria Pollutant Standards

The CAA requires the Environmental Protection Agency (EPA or the Agency) to establish National Ambient Air Quality Standards (NAAQS) (USC, § 7409) for certain pollutants, called "criteria pollutants," to protect the public health and welfare. Criteria air pollutants are:

- ozone, including precursors such as VOCs
- oxides of nitrogen (NO_x)
- carbon monoxide
- sulfur dioxide (SO_2)
- particulate matter (10 micrometers or smaller)
- lead

EPA develops air-quality control regions (USC, § 7407) and designates them as attainment or nonattainment areas for each of the criteria pollutants. For example, Boston and its suburbs are designated as an air-quality-control region that is in attainment for sulfur dioxide, but in nonattainment for carbon monoxide and ozone.

The air quality in attainment areas is at or better than the National Ambient Air Quality Standard (NAAQS) set by EPA; air quality in nonattainment areas is below the NAAQS. A source can determine the status of the area that it is located in either by consulting federal regulations (EPA, §§ 81.300–347) or by contacting the state or local air-pollution control agency.

State Implementation Plans

The CAA requires each state to adopt a plan, called a State Implementation Plan (SIP) (USC, § 7410), that imposes limits on emissions from air-pollution sources to the degree necessary to achieve and maintain the NAAQS in each air-quality-control region (or portion thereof) within the state. SIPs must assure attainment of the NAAQS by the date prescribed in the CAA. SIPs must meet federal requirements, but each state may choose its own mix of emissions controls for sources. SIP controls may include stationary and mobile source emissions limits; transportation plans; nonattainment-area (NAA) and prevention-of-significant-deterioration (PSD) permits for construction of new sources; and inspection and testing of vehicles and any other measures necessary for ensuring attainment or maintenance of the NAAQS. States can be more stringent than the federal requirements and have the primary role in enforcing SIPs. SIPS must be submitted to EPA for federal approval. Upon approval by EPA, a SIP can be enforced by the state or federal government.

Major Source Definitions

"Major" stationary sources are defined as sources that emit, or have the potential to emit, more than a prescribed amount of a designated pollutant. A "stationary source" is any source of air-pollution emissions, except for emissions from an internal combustion engine used for transportation. Many CAA requirements are keyed to whether a source is "major."

A "major source" can mean emitting or having the potential to emit as little as ten tons per year (TPY) or as much as 100 tons per year, depending on the pollutant emitted and the source's location. For example, for emissions of ozone precursors (VOCs, NO_x, and SO_2), the source is major if it is (USC, § 7511a):

located in	emits
an extreme area	10 tpy
a severe area	25 tpy
a serious area	50 tpy
a moderate/marginal	100 tpy

For emissions of carbon monoxide (CO), the source is major if it is (USC, § 7512a):

located in	emits
a serious area	50 tpy
a moderate area	100 tpy

For emissions of particulate matter (PM10), the source is major if it is (USC, § 7513a):

located in	emits
a serious area	70 tpy
a moderate area	100 tpy

For other criteria pollutants, sources emitting 100 tpy or more are major (USC, § 7602(j)). For hazardous air pollutants (HAP), which are addressed in the next section of this chapter, the definition of "major source" does not vary by pollutant or where the source is located. Any source that emits or has the potential to emit ten tons per year of any one, or 25 tons per year of any combination of, hazardous air pollutant is major (USC, § 7412(a)(1)).

Potential to Emit

In determining whether a source is major, and, therefore, subject to certain requirements, facility managers must consider not only actual emissions, but also a facility's potential to emit. Potential to emit is the maximum capacity of a stationary source to emit a pollutant under its physical and operational design. This implies that a source calculate emissions based on constant operation (24 hours per day, 365 days per year) at maximum capacity (all aspects of all operations), unless: (1) there is an inherent operational design limitation which precludes this, or (2) there is a federally enforceable permit provision which limits operation or requires pollution control equipment.

Many sources do not operate at their maximum potential-to-emit level for this duration, so their actual emissions are much less than their potential to emit. If a source takes no action, however, the Agency will impose requirements on the source based on its potential to emit.

If a source wishes to limit its obligations under the CAA, it can adopt certain conditions which limit its emissions to below the major source threshold. For example, a source could agree to limit its operations to 16 hours per day (e.g., two eight-hour shifts). These emissions limits must be "federally enforceable" to be considered. A federally enforceable limit must be directly enforceable by EPA or citizens and must be enforceable as a practical matter.

In recent guidance (EPA, 1995), EPA has clarified how a facility that emits hazardous or criteria air pollutants can avoid the application of certain requirements to its operations. Facilities emitting less than the threshold amounts are excluded from certain future controls and will not need to obtain an operating permit in the near term. The guidance also implements a transition policy that allows certain sources to be treated as minor during 1996 until they are able to obtain federally enforceable limitations.

The policy indicates that state and local agencies have a variety of vehicles for limiting a source's potential to emit (if submitted and approved by EPA). They include: federally enforceable state-operating permit programs (non-Title V), limitations established by rule, general permits, construction permits, Title-V permits, and source-specific SIP revisions. Practicable enforceability means that these vehicles contain accurate limitations and are specific regarding the sources covered, the time period for the limitation, and the method for determining compliance.

State and local authorities can create federally enforceable limitations for hazardous air pollutants as well. Because the definition of "potential to emit" in the National Emissions Standards for hazardous air pollutants regulations considers the effect of federally enforceable limitations on criteria pollutants on emissions of HAP, a source may not need to take a separate limit for HAP. For example, a limitation on the amount of a solvent used per month, which would limit the amount of VOC and HAP emissions from the solvent, might be sufficient to reduce a source's HAP emissions below the threshold level. If, however, the HAP is not a criteria pollutant, a state or local agency will need to seek approval from EPA for its HAP program pursuant to Section 112(l) of the CAA.

In this guidance, EPA established a transition policy which state or local agencies may take advantage of for up to two years from the date of the policy (until January 25, 1997). Sources maintaining emissions below 50% of all applicable major source requirements, sources with emissions above 50% with appropriate limits in state permits, and sources with emissions above 50% that emit noncriteria HAP and are subject to state restrictions will not be required to comply with major source requirements during that period. Such sources would not be treated as major and would not need to obtain a permit that limits their potential to emit.

Restrictions on Sources in Attainment Areas

States are required to adopt measures to prevent significant deterioration (PSD) of air quality (USC, §§ 7470-7479) in areas that are in attainment with the NAAQS. The PSD program is designed to allow industrial growth in attainment areas, while protecting air quality, through a construction permit system. PSD applies to new major stationary sources, certain modifications to major stationary sources, and modifications which are major stationary sources. Each new or modified emissions unit subject to PSD is required to undergo a best available control technology (BACT) review. BACT is defined (USC, § 7479(3)) as the emissions limitation, based on the maximum degree of reduction of each pollutant emitted from a major source, which the permitting authority—on a case-by-case basis, taking into account energy, environmental, and economic impacts and other costs—determines is achievable for such facility. The BACT analysis is performed for each pollutant subject to review emitted by the same emissions unit.

In 1987, the Agency established the Top-Down BACT method for determining BACT for a source. Under the Top-Down BACT analysis, all available technologies are ranked in descending order of effectiveness. The most stringent alternative is established as the standard unless the applicant shows, and the permitting authority agrees, that technical considerations or energy, environmental, or economic impacts indicate that the most stringent technology is not achievable.

For purposes of PSD, a major emitting facility (USC, § 7479) is a source in one of 26 designated categories that emits or has the potential to emit 100 tpy of any air pollutant. These sources are: (1) fossil-fuel fired steam electric plants of more than 250 million British thermal units per hour heat input, (2) fossil-fuel boilers of more than 250 million British thermal units per hour heat input, (3) petroleum storage and transfer facilities with a capacity exceeding 300,000 barrels, (4) municipal incinerators capable of charging more than 50 tons of refuse per day, (5) coal-cleaning plants (thermal dryers), (6) kraft pulp mills, (7) Portland Cement plants, (8) primary zinc smelters, (9) iron and steel mill plants, (10) hydrofluoric, sulfuric, and nitric acid plants, (11) petroleum refineries, (12) lime plants, (13) phosphate rock–processing plants, (14) coke-oven batteries, (15) sulfur recovery plants, (16) carbon black plants (furnace process), (17) primary lead smelters, (18) fuel-conversion plants, (19) sintering plants, (20) secondary metal–production facilities, (21) chemical-process plants, (22) taconite ore–processing facilities, (23) glass fiber–processing plants, (24) charcoal production facilities, (25) primary aluminum ore reduction plants, and (26) primary copper smelters. A source that is not within the 26 designated categories is a major source if it emits more than 250 tpy of any pollutant regulated under the CAA. The CAA excludes new or modified facilities that are nonprofit health or education institutions which have been exempted by the state from the definition of major emitting facility (USC, § 7479(1)). This exclusion may mean that such sources will not be subject to PSD or preconstruction permit requirements.

Any proposed major new source or major modification is subject to preconstruction review and permitting by EPA, by a state to whom the program is delegated, or by a state that has adopted PSD requirements in its SIP, to determine what emissions increases are allowed and to assure that they will not be exceeded (USC, § 7475). Where EPA has delegated such review, EPA and the state have concurrent enforcement authority. The permit describes the level of control to be applied and what portion of the increment may be made available to that source by the state. "Baseline" is the air quality that exists at the time the first application for a PSD permit is filed in a particular attainment area (USC, § 7479(4)). "Increments" are the maximum amount of deterioration that can occur in an attainment area over the baseline (USC, § 7473).

Restrictions on Sources in Nonattainment Areas

The CAA classifies areas as nonattainment (USC § 7502) for the type of pollutant which is found in excess of one of the NAAQS (*e.g.*, ozone) and for the
degree to which it exceeds the NAAQS (*e.g.*, extreme, severe, serious, moderate, marginal). For example, Denver-Boulder is a serious CO nonattainment
area. The CAA establishes deadlines for attaining the NAAQS which vary
depending on classification of the area. States that contain nonattainment areas
must impose reasonably available control technology (RACT) on existing
sources. RACT is the lowest emissions limit that an existing source can meet by
application of control technology that is reasonably available and economically
feasible.

States are also required to prevent further deterioration of air quality in
nonattainment areas by requiring permits for the construction and operation of
new and modified stationary sources in nonattainment areas. The nonattainment
area (NAA) new source review program is designed to allow industrial growth in
nonattainment areas (if certain stringent requirements are met), while preventing
deterioration of air quality. The NAA permit program applies to new major stationary sources and to major modifications. The NAA permit program requires
sources wishing to construct or modify to achieve the lowest achievable emission
rate (LAER) (USC, § 7503). LAER is defined as the rate of emissions that
reflects either the most stringent emissions limitation that is contained in the SIP
of any state for a source type (unless the owner or operator demonstrates that such
limitations are not achievable) or the most stringent limitation that is achieved in
practice by a source type, whichever is more stringent (USC, § 7501). The NAA
permit program requires that emissions be offset (it permits no net emissions
increase). The CAA imposes additional requirements on areas that are nonattainment for ozone (USC, §§ 7511-7511f), carbon monoxide (USC, §§ 7512 and
7512a), particulate matter (USC, §§ 7513-7513b), sulfur oxides, nitrogen dioxide, and lead. The requirements imposed on sources increase in stringency with
the degree to which the area exceeds the NAAQS.

5.4   REQUIREMENTS BASED ON ACTIVITY
OR POLLUTANT EMITTED

In addition to the SIP regulatory scheme, the CAA establishes two other major
regulatory programs for stationary sources that apply to certain types of sources
regardless of where they are located: New Source Performance Standards and
National Emissions Standards for Hazardous Air Pollutants.

New Source Performance Standards

The New Source Performance Standards (NSPS) (USC, § 7411) program establishes stringent emissions limitations for specific pollutants from sources in designated industrial categories regardless of the geographic area in which the source is located or the air quality associated with the area. To date, the Agency has promulgated more than 75 Standards of Performance. Among them are rules that regulate small industrial, commercial, and institutional steam-generating units, incinerators, grain elevators, and rotogravure printing presses.

These standards apply to both new and modified sources. In some cases, existing facilities are affected. "New" means constructed on or after the proposal date of the regulation (USC, § 7411(a)(2)). A modification is any change that results in an increase of emissions to the atmosphere, with some exemptions (USC, § 7411(a)(4)). NSPS is a federally developed and promulgated program, but many states are delegated authority to implement and enforce. EPA, however, retains authority to enforce.

Medical-Waste Incinerators

Of particular concern to hospitals is the NSPS for medical waste incinerators which is a source category specifically required to be regulated by a NSPS under the CAA (USC, § 7429). These standards must reflect the maximum degree of reduction in emissions of air pollution, taking into consideration cost and non–air quality health and environmental impacts and energy requirements achievable for new or existing units. For new units, the standard may be no less stringent than the emissions limit achieved by the best-controlled source in the category. For existing units, the standard may be no less stringent than the average emissions of the best-performing 12% of units in the source category.

For medical-waste incinerators, the CAA requires the Agency to set an emissions limit for particulate matter, opacity (only if appropriate), sulfur dioxide, hydrogen chloride, oxides of nitrogen, carbon monoxide, lead, cadmium, mercury, dioxins, and dibenzofurans. The CAA does not permit the Agency to require a particular technological system of continuous emissions reduction; the Agency may only set emissions limits.

The Agency recently proposed a rule to implement this provision (Federal Register, February 27, 1995) and is under a court order to finalize a rule applicable to this source category by April 15, 1996. The proposal set stringent emissions limits for the regulated pollutants, including a zero-percent opacity limit for fugitive emissions of fly ash and bottom ash. The standards set apply to all medical-waste incinerators regardless of size or type or whether the source is existing or new. Many of the standards appear to be achievable only with the use of one technology, a fabric filter system with carbon injection. The regulation also proposes rigorous operator training, inspection, monitoring, and record-keeping requirements.

Hazardous Air Pollutants

Hazardous air pollutants include the 189 listed substances, as well as any additional substances which the Administrator determines present or may present a threat of adverse human health effects, including substances that may reasonably be anticipated to be carcinogenic, mutagenic, teratogenic, neurotoxic; to cause reproductive dysfunction; or are acutely or chronically toxic or a threat of adverse environmental effects including ambient concentrations, bioaccumulation, or deposition (USC, § 7412(a)(6)).

The Clean Air Act, as amended in 1990, directs the Agency to regulate hazardous air pollutants, not by pollutant, but by source category (USC, § 7412(c)). In other words, the Agency does not issue a NESHAP for chromium; it issues a NESHAP for magnetic tape production because that production process emits chromium and other hazardous air pollutants. Dry cleaners, wood-furniture manufacturers, and synthetic organic chemical manufacturers are just a few of the sectors covered by NESHAPs. Potentially of concern to hospitals and universities is the NESHAP for demolition and renovation of facilities that contain asbestos-containing material (EPA, Part 61, Subpart M). This NESHAP establishes work-practice standards for anyone removing, handling, and disposing of asbestos-containing material. Hospitals and universities that are considered to be employers of state or local government employees may be subject to additional regulation for asbestos removal under the Asbestos Hazard Emergency Response provisions of the Toxic Substances Control Act (EPA, § 763.120–121).

The NESHAP program requires application of demonstrated technology-based emissions standards, called maximum available control technology (MACT). MACT is defined for new sources as the best control in use on a similar source. MACT for existing sources is the average emissions limit for the best 12% of existing sources *or* if less than 30 sources exist in the category, the average emissions limit of the best five sources (USC, § 7412(d)(3)). NESHAPs apply to major sources (existing, new, modified, or reconstructed) and area sources (existing and new). For area sources, the Agency may set a standard based on generally available control technology (GACT) which is a less-stringent standard than MACT. An area source is a source that is not major, for example, a dry cleaners. If a technology-based emissions standard is not feasible, the Agency can set a work practice or operational standard.

A new source is any source that commenced construction or reconstruction after a MACT standard is proposed. New sources must comply upon promulgation of the MACT rule or at the initial start-up of the business, whichever is later. The CAA gives existing sources up to three years to comply with a MACT after promulgation and six *additional* years to comply if the source makes an early reduction (90%) of its emissions. The NESHAP program is to be implemented through operating permits.

The CAA also requires the Agency to issue standards to prevent accidental release of hazardous air pollutants (USC, § 7412(r)). EPA has issued an initial list of 100 substances of particular concern, listing threshold quantities for each substance (Federal Register, January 31, 1994) and proposed requirements for risk-management programs which would outline procedures to prevent and mitigate accidental releases (Federal Register, October 20, 1993).

5.5   ELECTRIC GENERATORS
AND ACID-DEPOSITION CONTROL

Congress authorized EPA to establish the Acid Deposition Program with the goal of reducing sulfur dioxide (SO_2) and nitrogen oxide (NO_x) emissions. Acid deposition occurs when sulfur dioxide and nitrogen oxide emissions are transformed in the atmosphere and return to earth as acidic rain, fog, or snow. To achieve reductions, the program employs both traditional and innovative market-based approaches for controlling air pollution. A facility that cogenerates steam and electricity is not a utility for the purposes of this program unless the unit was constructed for the purpose of supplying, or commenced construction after November 15, 1990 and supplies, more than one-third of its electrical output capacity and more than 25 megawatts electrical output for sale (USC, § 7651a(17)(C)).

The CAA sets as its primary goal the reduction of annual SO_2 emissions by ten million tons below 1980 levels. To achieve these SO_2 reductions, EPA is allocating allowances (USC, § 7651b) to sources in two phases (USC, §§ 7651c and 7651d). Each allowance will permit a utility to emit one ton of SO_2 (USC, § 7651a(3)). Sources are allocated allowances based on required emissions rates and baseline fuel consumption.

Phase I began on January 1, 1995, and affects 110 mostly coal-burning electric utility plants (100 megawatts or greater, that emit more than a certain amount of SO_2). By that date, affected sources were required to achieve certain SO_2 reductions. Phase II, which begins in the year 2000, tightens the annual emissions limits imposed on these larger, higher-emitting plants and also sets restrictions on smaller, cleaner plants fired by coal, oil, and gas (approximately 2,000 utilities) (USC, § 7651d). All existing utility units with an output capacity of 25 megawatts or greater and all new utility units will be affected by Phase II. In both phases, affected sources are required to monitor their emissions continuously in order to assure compliance.

Utilities may freely trade allowances within their systems and/or buy or sell to or from other sources. Each source must have sufficient allowances to cover its annual emissions. The Agency has set up an allowance tracking system to issue, record, and track allowances.

The CAA also requires that these sources reduce their emissions of nitrogen oxides (USC, § 7651f). The Agency issued regulations imposing emissions limits for utility boilers using low NO_x burner technology (Federal Register, Mar. 22, 1994). That rule was challenged, and now EPA is required to repropose the rule. The CAA also requires the Agency to revise its NSPS applicable to fossil-fuel steam-generating units to impose restrictions which further reduce NO_x missions (USC, § 7651f(c)).

5.6   RESTRICTIONS ON REFRIGERANTS AND OTHER STRATOSPHERIC OZONE-DEPLETING SUBSTANCES

Both international agreements and the CAA require a complete phaseout of certain ozone-depleting substances: halons, chlorofluorocarbons (CFCs), carbon tetrachloride, methyl chloroform, methyl bromide, and hydrochlorofluorocarbons (HCFCs). Halons were phased out on January 1, 1994; chlorofluorocarbons, methyl chloroform, and carbon tetrachloride were phased out on January 1, 1996. Methyl bromide will be phased out in 2001. The phaseout of HCFCs will begin in 2003. All HCFCs will be phased out by 2030. The phaseout is implemented through an allowance system, except for HCFCs. All persons who wish to import or produce ozone-depleting substances must possess allowances prior to importation or production (USC, §§ 7671c and 7671d).

The CAA also requires that EPA publish a list of safe and unsafe substitutes for ozone-depleting substances and to ban the use of substitutes that the Administrator has determined may present adverse effects to human health or the environment where the Administrator has identified an alternative that reduces overall risk and is currently or potentially available (USC, § 7671k).

The CAA also imposes restrictions on the service and disposal of appliances (USC, § 7671g) and motor vehicles (USC, § 7671h) containing ozone-depleting substances. All technicians who service, repair, or dispose of appliances or motor vehicles must be certified by an EPA-approved program. The sale of refrigerant is limited to certified technicians, with certain exceptions. EPA has promulgated regulations to implement these provisions, imposing work-practice and equipment standards on those who service or dispose of appliances or motor vehicles (EPA, Part 82, Subparts B and F). Universities, schools, and hospitals must take care to ensure that any maintenance or activity be performed by certified technicians and that appropriate recycling guidelines are followed. Indiscriminate release of CFCs from existing units can result in fines.

The CAA prohibits any person from selling or distributing or offering for sale or distribution any product that releases ozone-depleting substances identified either in the CAA or by the Administrator as nonessential (USC, § 7671i). In addition, all containers of certain ozone-depleting substances or products containing or manufactured with certain ozone-depleting substances must bear a warning label (USC, § 7671j).

5.7   STANDARDS FOR MOTOR VEHICLES AND FUELS

Up to this point, we have been addressing emissions from stationary sources. The Clean Air Act also governs emissions from mobile sources. Most people are familiar with the passenger-car standards that motor-vehicle manufacturers face for average fuel-economy and tailpipe emissions.

The 1990 Amendments to the Clean Air Act continue these standards and require EPA to establish more stringent federal emissions standards applicable to new motor vehicles (USC, § 7521) and further regulate fuels used in motor vehicles (USC, § 7545). In addition, the 1990 Amendments mandate that a certain percentage of fleet vehicles use "clean" fuels (USC, § 7586). This places a new obligation on educational or health-care facilities that own ten or more vehicles and are located in 22 specified urban areas. If a facility is subject to these provisions, beginning with the 1998 model year, a certain percentage of vehicle purchases must meet clean-fuel vehicle (CFV) criteria.

Even if your facility is not located in a "covered" urban area, please note that EPA expects that the CFF will eventually be expanded to include the entire country. Thus, it may be wise to familiarize yourself with the program's requirements and opportunities as part of your long-range planning efforts, even if your location is not currently "covered."

Another provision of the CAA prohibits anyone from removing or rendering inoperative emissions-control devices or elements of design either prior to or after sale and delivery of the vehicle to the purchaser. For example, using leaded fuel in a vehicle requiring unleaded fuel or removing the catalytic converter from a vehicle is prohibited. Both acts are deemed to render the emissions-control system inoperative.

Fleet Vehicle Owners and Operators

The Clean Fuel Fleet (CFF) Program sets forth new regulatory requirements for owners or operators of "covered fleets" in approximately two dozen "covered areas" around the country. These covered areas are metropolitan regions that have experienced particularly poor air quality over the last decade. Beginning with the 1998 model year, owners and operators of covered fleets that are located in covered areas (and whose states have not opted out of the CFF and developed their own programs) will have to ensure that a certain percentage of their new vehicles meet clean-fuel vehicle (CFV) emission standards. This percentage can be achieved by any combination of the following:

1. converting existing conventional vehicles to CFVs
2. purchasing or leasing new CFVs
3. purchasing CFF program emission "credits"

Each state will incorporate its own version of the CFF emissions-credit program into its State Implementation Program (SIP).

Does the CFF Program Apply to My Facility?

The CFF program is a state-run program. Although certain aspects of the program are mandatory and EPA has issued program guidelines, each state will have a good deal of latitude in deciding how to structure its CFF program, including determinations of what constitutes a "centrally fueled" fleet and "operating in a covered area." Thus, health and education facilities should be sure to contact their state air-pollution-control agency for details on how to comply with its version of the CFF program. Leaving aside for the moment these state-by-state variations, the CFF program will apply to a facility *only* if:

1. it is located in a "covered" area
2. its fleet meets the definition of a "covered" fleet

A covered area is any one of the metropolitan areas listed in Table 5–1. They are known as ozone and/or carbon monoxide "nonattainment areas" because they have been unable to achieve the level of air quality set for them by EPA.

A "covered fleet" is a group of ten or more motor vehicles owned, operated, leased, or otherwise controlled by a single person (a person can be an individual, a private business, or federal, state, or local government), which is *capable* of being centrally fueled (USC, § 7581 and Federal Register, December 9, 1993). This includes all sizes of cars, and trucks and buses with a gross vehicle weight of up to 26,000 pounds. It does not, however, include:

a. vehicles that are parked at personal residences each night under normal circumstances
b. vehicles that cannot be fueled at a central location

Table 5–1.   States and Areas Affected by the Clean-Fuel Fleet Program

Affected Area:	State(s):
1. Atlanta	Georgia
2. Baltimore	Maryland
3. Baton Rouge	Louisiana
4. Beaumont-Port Arthur	Texas
5. Boston-Lawrence-Worcester (Eastern Mass.)	Massachusetts, New Hampshire
6. Chicago-Gary-Lake County	Illinois, Indiana
7. Denver-Boulder	Colorado
8. El Paso	Texas
9. Greater Connecticut	Connecticut
10. Houston-Galveston-Brazoria	Texas
11. Los Angeles-South Coast Air Basin	California
12. Milwaukee-Racine	Wisconsin
13. New York-Northern New Jersey-Long Island	Connecticut, New Jersey, New York
14. Philadelphia-Wilmington-Trenton	Delaware, Maryland, New Jersey, Pennsylvania
15. Providence (All Rhode Island)	Rhode Island
16. Sacramento Metro	California
17. San Diego	California
18. San Joaquin Valley	California
19. Southeast Desert Modified AQMA	California
20. Springfield (Western Mass.)	Massachusetts
21. Ventura County	California
22. Washington (District of Columbia)	Maryland, Virginia, District of Columbia

In addition to the exemptions above for specific vehicle types, certain types of vehicle fleets which would otherwise be "covered" fleets have been exempted, and *do not* need to comply with the CFF regulations:

   a. retail rental vehicles
   b. emergency vehicles
   c. law-enforcement vehicles
   d. nonroad vehicles

If you have any questions about whether your location, your fleet, or a particular vehicle falls within these definitions, you should consult with your state air-pollution-control agency or the appropriate EPA regional office.

CFF Program Requirements

The CFF program requirements go into effect for the 1998 model year. Thus, though they may choose to do so, covered fleet owners and operators are not required to make any changes in their purchasing or conversion plans until that time. The percentage of each fleet's new acquisitions which must be made up of CFVs will increase from 1998 to 2000 as follows (USC, § 7586):

Light-Duty Vehicles and Trucks
1998—30% of all new acquisitions
1999—50% of all new acquisitions
2000—70% of all new acquisitions

Heavy Duty Vehicles
Beginning in 1998, 50% of all new acquisitions

A CFV is one that is capable of running on one of EPA's recommended "alternative fuels," including: methanol, ethanol, mixtures of alcohols with gasoline or other fuels, reformulated gasoline, diesel, natural gas, liquefied petroleum gas, hydrogen, and electric cell. Almost any alternative fuel is permitted, so long as the vehicle's emissions comply with CFV standards.

The CFF's CFV standards are essentially identical to those of the California Air Resources Board's (CARB) Low-Emission Vehicle (LEV) Program. In order to qualify as part of a covered fleet CFF compliance program, a CFV must meet one of the three emissions classifications below:

1. a low-emission vehicle (LEV)
2. an ultra-low-emission vehicle (ULEV)
3. a zero-emission vehicle (ZEV)

Obviously, to qualify as part of a CFF compliance program, a fleet's CFVs must use clean alternative fuels, as defined in Section 241(2) of the CAA, when operating in covered areas.

The CFF program standards are expected to reduce emissions by approximately two-thirds in comparison to conventional vehicles for cars and light trucks. Slightly less-rigorous requirements will be set for heavy-duty trucks.

The CFF program divides CFVs into three weight classes:

1. *Light-duty vehicles (LDVs):* Up to 8,500 pounds GVWR.[1] In order for a LDV to qualify as a CFV, it must meet the emission standards set forth in Sections 242 and 243 of the CAA.
2. *Light-duty trucks (LDTs):* Up to 8,500 pounds GVWR. In order for a LDT to qualify as a CFV, it must meet the emission standards set forth in Sections 242 and 243 of the CAA.
3. *Heavy-duty vehicles (HDVs):* Between 8,500 pounds and 26,000 pounds. In order for an HDV to qualify as a CFV, it must meet the emission standards set forth in Sections 242 and 245 of the CAA (Federal Register, September 30, 1994).

EPA will require vehicle manufacturers to label LEVs, ULEVs, and ZEVs, so that purchasers (e.g., covered fleet owners and operators) will know which vehicles will qualify for the CFF program.

Conversion

As an alternative to purchasing or leasing new CFVs, covered fleet owners and operators may meet their required CFF program new vehicle percentage by converting some of their current conventional vehicles into CFVs. These converted vehicles will be eligible for the same transportation-control measure (TCM) and emissions credit provisions extended to newly purchased CFVs, so long as the vehicle is converted in accordance with EPA's regulations and meets the required emissions-performance criteria for its vehicle class (EPA, §§ 242-245).

All CFV conversions will need to meet the emission standards set forth in 40 C.F.R. Part 88 for their vehicle class (i.e., LDV, LDT, or HDV) and conversion rating (i.e., LEV, ULEV, or ZEV). After the conversion, the vehicle must be able to operate on a fuel other than the one for which it was originally manufactured. Converted CFVs must be labeled as such on their engine labels. The original equipment manufacturer (OEM) remains responsible for original equipment, unless the conversion causes the equipment to malfunction or fail.

Benefits of Compliance

Those fleet owners who are in compliance with the CFF program will be eligible to earn CFF emissions credits, which they can then sell to other fleet owners who are not in compliance with the program. In addition, fleet owners and operators who begin purchasing CFVs prior to 1998, who purchase greater than the

---

[1] Gross Vehicle Weight Rating.

required percentage of new CFVs, or whose fleets include ULEV or ZEV vehi-cles will also be eligible for CFF vehicle purchase credits (USC, § 7586(f)).

Moreover, some states may choose to promote CFVs and alternative fuels by charging higher registration fees for non-CFVs, and by offering positive incen-tives (e.g., preferred parking plans, high-occupancy vehicles (HOV) exemptions) to complying covered fleet owners and operators. Complying fleet owners/oper-ators will also be exempted from certain transportation-control measures (Federal Register, March 1, 1993).

Additional Aspects of the Program

The CFF directs automobile manufacturers to begin marketing CFVs able to run on "clean fuels" (e.g., alcohol, electricity) and requires the working life span of pollution-control devices to be increased from 50,000 to 100,000 miles. It also requires petroleum refiners to begin offering reformulated gasoline—"clean fuels"—for sale as of January 1, 1995, in the nine metropolitan areas with the highest ozone levels. Vapor-recovery nozzles are now required in these same areas, and efficient commuting (e.g., van pools, HOV lanes) must be encouraged by local agencies. In colder areas prone to carbon monoxide (CO) nonattainment, petroleum refiners must offer oxygenated fuel for sale. Finally, though presently unaffected, the CAA amendments direct EPA—if it finds it to be necessary—to regulate nonroad engines and vehicles (e.g., locomotives, construction equip-ment, lawn/yard equipment) as well. In doing so, the Agency will consider the CFF standards currently being applied to on-road vehicles.

The Energy Policy Act Alternative Fuel Vehicle Program

Apart from and in addition to the CFF program (which is managed by EPA), the Department of Energy (DOE) administers the Energy Policy Act of 1992 (EPACT) program, which is designed to promote the use of alternative fuels. EPACT has its own fleet vehicle program, applying only to vehicles with a gross vehicle weight (GVW) of 8,500 pounds or less. It requires certain fleets of 20 or more vehicles located in any of 125 consolidated metropolitan statistical areas (CMSAs) throughout the country to ensure that a set percentage of their newly purchased vehicles are alternative-fuel vehicles (AFVs). Thus, while the EPACT program is more limited than the CFF program in terms of fleet size and vehicle type, its geographic applicability is much broader.

The EPACT program already requires the federal government to purchase a certain number of AFVs each year. Beginning with the 1996 calendar year, state entities and "alternative fuel providers" are also subject to the EPACT requirements. The DOE has proposed a rule to implement these requirements, but it is not final as of this writing (Federal Register, February 28, 1995). It is, therefore, not clear how the term "state entity" will be defined. In addition, if current program incentives fail to generate sufficient voluntary interest in the AFV program, the DOE is authorized to extend the its regulations to private and municipal fleets beginning in 1999.

A wide variety of alternative fuels will qualify a vehicle as an AFV, including methanol, natural gas, hydrogen, and electricity. AFVs are expected to be exempt from many pending transportation-control measures (e.g., HOV lanes, time-restricted use, employee trip-reduction programs). In addition, tax credits, tax deductions, and purchase credits are available to all owners and operators of qualifying AFVs, even those who are not required by the law to have these fleets. As with the CFF program, the percentage of new vehicles within each fleet that must be AFVs begins to increase once the program takes effect.

5.8 EMPLOYEE COMMUTE OPTIONS PROGRAM

EPA administers the Employee Commute Options (ECO) Program, often referred to as the Employee Trip Reduction (ETR) Program. This program is designed to decrease air pollution by reducing solo driving and encouraging the use of car pools and public or alternative transportation. At present, it applies to employers with 100 or more employees at a facility that is located in a severe or extreme ozone nonattainment or serious carbon-monoxide nonattainment area (see Table 5–2). Contrary to many reports, the ECO program will not require employees to change their commuting habits, nor will it penalize employers who make good-faith efforts to comply with its requirements.

Owners and operators of such facilities must determine the average passenger occupancy (APO) of their employees' commute. They must then submit a proposed compliance program to the state air-protection program for review. This program should ensure that the facility's APO will be at or above the target APO. ECO programs can include measures such as mass-transit vouchers, subsidized carpooling, bicycling, telecommuting from home, and compressed workweeks.

Employers with an APO already at or above the target APO will not need to take any further action. In addition, many states have created a "compliance exemption" for employers who do not reach the target APO, but can demonstrate good-faith efforts at compliance. The structure and details of these programs must first be written into each state's State Implementation Plan (SIP); affected employers should thus be sure to contact their state air-pollution-control agency if required to design a Trip Reduction Program.

Table 5–2.   States and Areas Affected by the Employee Commute Options Program

Affected Area:	State(s):
1. Baltimore	Maryland
2. Chicago-Gary-Lake County	Illinois, Indiana
3. Houston-Galveston-Brazoria	Texas
4. Los Angeles-South Coast Air Basin	California
5. Milwaukee-Racine	Wisconsin
6. New York-Northern New Jersey-Long Island	Connecticut, New Jersey, New York
7. Philadelphia-Wilmington-Trenton	Delaware, Maryland, New Jersey, Pennsylvania
8. Southeast Desert Modified AQMA	California
9. Ventura County	California

5.9   OTHER AIR ISSUES

The Clean Air Act requires that EPA impose "enhanced monitoring" on sources and require submission of compliance certifications (USC, § 7414(a)(3)). Compliance certifications must identify the applicable requirement that the certification is based on, the method used for determining compliance, the compliance status of the institution, and whether compliance is continuous or intermittent (USC, § 7414(a)(3)). The CAA requires that EPA promulgate rules to implement this provision. While EPA has issued a highly contested proposal, it has been unable to finalize regulations, despite a court order to do so (Federal Register, Oct. 22, 1993). The CAA does, however, contain a general authority which permits EPA to require sources subject to any of the CAA's requirements to "install, use, and maintain . . . monitoring equipment" (USC, § 7414(a)(1)(C)).

Most states and many local governments have laws that regulate odors from facilities. These restrictions often are found in the state or local general nuisance statutes and regulations. A facility can be fined for emitting an odor if it may be dangerous or may adversely affect public health. States usually exempt agricultural activities from such restrictions. As a practical matter, this is usually a question of being a good neighbor and minimizing nuisance odor. For operations and activities that unavoidably generate noticeable odor, facility managers may wish to consider a community outreach program to educate their neighbors on the source of the odor and, hopefully, the reasons for its being benign.

Finally, while federal and state officials can take enforcement action, note that the CAA also allows citizens to commence civil actions against any person for violating the Clean Air Act or federal or state administrative orders (USC,

§ 7604). In addition, a citizen may sue EPA when it fails to perform a nondiscretionary duty. Finally, a citizen may sue any person who proposes to construct or constructs a new or modified source without an appropriate permit or is in violation of such a permit.

REFERENCES

(CFR in text means Code of Federal Regulations)
California Administrative Code, Title 13.
40 Code of Federal Regulations §§ 81.300–437 (1994).
40 Code of Federal Regulations § 88.101–94
40 Code of Federal Regulations §§ 242–245.
40 Code of Federal Regulations § 763.120–121.
40 Code of Federal Regulations Part 61, Subpart M.
40 Code of Federal Regulations Part 82, Subparts B and F.
58 Federal Register 11,888 (March 1, 1993).
58 Federal Register 54,190 (October 20, 1993).
58 Federal Register 54,648 (October 22, 1993).
58 Federal Register 64,679 (December 9, 1993).
59 Federal Register 4478 (January 31, 1994).
59 Federal Register 13,538 (March 22, 1994).
59 Federal Register 50,042 (September 30, 1994).
60 Federal Register 10,654 (February 27, 1995).
60 Federal Register 10,970 (February 28, 1995).

O'Reilly, James, William Luneburg, Kim Burke, and Richard Ayres. 1995. *Clean Air Permitting Manual.* Colorado Springs: Shepards/McGraw-Hill.
U.S. EPA. 1995. Options for Limiting the Potential to Emit (PTE) of a Stationary Source under Section 112 and Title V of the Clean Air Act, Memorandum (January 25, 1995).
U.S. EPA. 1994. Proposed Rules. Federal Register 59:44460 (Aug. 29, 1994).

42 United States Code §§ 7401 *et seq.*
42 United States Code § 7407.
Section 109 of the Clean Air Act, 42 United States Code § 7409.
Section 110 of the Clean Air Act, 42 United States Code § 7410.
Section 111 of the Clean Air Act, 42 United States Code § 7411.
Section 111(a)(2) of the Clean Air Act, 42 United States Code § 7411(a)(2).
Section 111(a)(4) of the Clean Air Act, 42 United States Code § 7411(a)(4).
Section 112(a)(1) of the Clean Air Act, 42 United States Code § 7412(a)(1).

Section 112(a)(6) of the Clean Air Act, 42 United States Code § 7412(a)(6).
Section 112(c) of the Clean Air Act, 42 United States Code § 7412(c).
Section 112(d)(3) of the Clean Air Act, 42 United States Code § 7412(d)(3).
Section 112(r) of the Clean Air Act, 42 United States Code § 7412(r).
Section 114(a)(3) of the Clean Air Act, 42 United States Code § 7414(a)(3).
Section 114(1)(1)(C) of the Clean Air Act, 42 United States Code § 7414 (a)(1)(C).
Section 129 of the Clean Air Act, 42 United States Code § 7429.
Sections 160–169 of the Clean Air Act, 42 United States Code §§ 7470–7479.
Section 163 of the Clean Air Act, 42 United States Code § 7473.
Section 165 of the Clean Air Act, 42 United States Code § 7475.
Section 169 of the Clean Air Act, 42 United States Code § 7479.
Section 169(1) of the Clean Air Act, 42 United States Code § 7479(1).
Section 169(3) of the Clean Air Act, 42 United States Code § 7479(3).
Section 169(4) of the Clean Air Act, 42 United States Code § 7479(4).
Section 171 of the Clean Air Act, 42 United States Code § 7501.
Section 172 of the Clean Air Act, 42 United States Code § 7502.
Section 173 of the Clean Air Act, 42 United States Code § 7503.
Sections 181 through 185B of the Clean Air Act, 42 United States Code § 7511–7511f.
Section 182 of the Clean Air Act, 42 United States Code § 7511a.
Sections 186 and 187 of the Clean Air Act, 42 United States Code §§ 7512 and 7512a.
Section 187 of the Clean Air Act, 42 United States Code § 7512a.
Sections 188 through 190 of the Clean Air Act, 42 United States Code §§ 7513–7513b.
Section 189 of the Clean Air Act, 42 United States Code § 7513a.
Section 202 of the Clean Air Act, 42 United States Code § 7521.
Section 211 of the Clean Air Act, 42 United States Code § 7545.
Section 241 of the Clean Air Act, 42 United States Code § 7581.
Section 246 of the Clean Air Act, 42 United States Code § 7586.
Section 302(j) of the Clean Air Act, 42 United States Code § 7602(j).
Section 304 of the Clean Air Act, 42 United States Code § 7604.
Section 402(3) of the Clean Air Act, 42 United States Code § 7651a(3).
Section 402(17)(C) of the Clean Air Act, 42 United States Code § 7651a(17)(C).
Section 403 of the Clean Air Act, 42 United States Code § 7651b.
Section 404 and 405 of the Clean Air Act, 42 United States Code §§ 7651c and 7651d.
Section 407 of the Clean Air Act, 42 United States Code § 7651f.
Section 407(c) of the Clean Air Act, 42 United States Code § 7651f(c).
Section 501(2) of the Clean Air Act, 42 United States Code § 7661(2).
Section 502(b) of the Clean Air Act, 42 United States Code § 7661a(b).
Section 502(b)(9) of the Clean Air Act, 42 United States Code § 7661(b)(9).

Section 503 of the Clean Air Act, 42 United States Code § 7661b.
Section 503(b)(2) of the Clean Air Act, 42 United States Code § 7661(b)(2).
Section 504 of the Clean Air Act, 42 United States Code § 7661c.
Section 504(f) of the Clean Air Act, 42 United States Code § 7661c(f).
Sections 604 and 606 of the Clean Air Act, 42 United States Code §§ 7671c and 7671e.
Section 608 of the Clean Air Act, 42 United States Code § 7671g.
Section 609 of the Clean Air Act, 42 United States Code § 7671h.
Section 610 of the Clean Air Act, 42 United States Code § 7671i.
Section 611 of the Clean Air Act, 42 United States Code § 7671j.
Section 612 of the Clean Air Act, 42 United States Code § 7671k.

CHAPTER *6*

Wastewater and Storm-Water Issues

6.1 REGULATORY FRAMEWORK

Water pollution in the United States is regulated primarily under the Federal Water Pollution Control Act (FWPCA), more commonly known as the Clean Water Act (CWA) (USC, § 1251 et seq.). The objective of the CWA is to "restore, maintain, and preserve the integrity of the nation's waters" by restricting the discharge of pollutants. The CWA is administered by the U.S. Environmental Protection Agency (EPA), although a substantial amount of authority under the CWA has been delegated to the states (Menell and Stewart 1994). In addition to administering the Federal CWA program within their boundaries, many states have also enacted their own water-quality laws.

The CWA subdivides water-pollution sources into two fundamental categories: (1) point sources and (2) nonpoint sources. Universities, hospitals, and schools are usually subject to regulation as point sources, but may have features that make them nonpoint sources, as well.

Point Sources

Point sources are defined as "discernible, defined, discrete conveyances . . . from which pollutants are or may be discharged" (USC, § 1362(14)). Almost any discharge can be a point source—everything from a spout, pipe, or ditch to machinery and waterborne vessels. Some sources that for practical purposes are nonpoint sources are regulated under the CWA as point sources. For example, a concentrated animal-feeding operation (CAFO) is considered a point source. The term "point source" is defined in the CWA as any discernible, confined and discrete conveyance, including but not limited to any pipe, ditch, channel, tunnel, conduit, well, discrete fissure, container, rolling stock, concentrated animal feeding operation, or vessel or other floating craft, from which pollutants are or may be discharged. This term does not include agricultural storm-water discharges and return flows from irrigated agriculture.

Use of the phrase "may be discharged" in the definition of "point source" is significant because it requires permits for sites (such as surface impoundments) that do not normally discharge pollutants, but which may do so under unusual circumstances, such as excessive rainfall.

Direct Dischargers

The CWA further subdivides point sources into either (1) direct dischargers or (2) indirect dischargers. A person who intends to discharge pollutants from a point source directly into a body of water is categorized as a "direct discharger." Direct dischargers must obtain a National Pollutant Discharge Elimination System (NPDES) permit prior to such discharge and may have to install and use pollution-control equipment. The type of equipment required depends on the type of pollutants discharged (e.g., solids, grease, biological waste, toxics), and on whether the point source is a "new" or "existing" source.

Indirect Dischargers

In contrast to a direct discharger, a person intending to discharge pollutants from a point source into a publicly owned wastewater treatment (POTW) system is considered an "indirect discharger." Indirect dischargers must ensure that their pollutant discharges do not pass through or interfere with the operation of the POTW that is downstream from them. This may require them to pretreat their discharges.

Nonpoint Sources

A nonpoint source (NPS) is any source of pollutants that is *not* a discernible, defined, discrete conveyance (e.g., runoff from automobile parking lots). By statutory definition, NPSs include certain point sources (e.g., irrigation return flows), as well. While NPS pollution is addressed in Section 319 of the CWA, which encourages area-wide waste-treatment management by the states, it is virtually unregulated under the CWA. Nevertheless, EPA is authorized to promulgate regulations governing a facility's site runoff, drainage, and waste disposal if it chooses to do so.

6.2   DIRECT DISCHARGER PERMITS

Activities Subject to NPDES-Permitting Requirements

Simply put, if your institution directly discharges wastewater into a body of water, you are probably subject to regulation under the National Pollutant Discharge Elimination System (NPDES), set forth at Section 402 of the CWA. While most schools and hospitals are not direct dischargers, and instead use sewer lines to a publicly owned treatment works, the following discussion is included for those facilities that do qualify, and to provide a thorough discussion of this area for all facility managers.

Under the CWA, the direct "discharge of a pollutant" by any person is unlawful unless the discharge is in compliance with an NPDES permit. Although permits are most frequently issued to dischargers on an individual or site-by-site basis, EPA is authorized to issue "general permits" for categories of discharges as necessary. Permits are issued either by EPA or by a state that has been delegated permitting authority under Section 402(b) of the CWA. Permits are issued for a fixed term not to exceed five years. Forty states are currently authorized to administer their own NPDES programs. State programs must be at least as stringent as the federal program. States seeking to administer their own programs are required to enter into a Memorandum of Agreement with EPA describing terms, conditions, and agreements relevant to the administration and enforcement of the state program (EPA CFR § 123.24).

It is important to note that not all pollutant discharges are subject to the NPDES permit program. The NPDES program requires permits only for the discharge of "pollutants" from any "point source" into the "waters of the United States."

- The term "pollutant" is defined to include a variety of substances such as solid waste, incinerator residue, sewage, garbage, chemical wastes, biological and radioactive materials, heat, rock, and industrial, municipal, and agricultural waste. Courts have interpreted the term "pollutant" broadly enough to include practically all waste material. For example, bombs dropped on naval target ranges have been held to be "pollutants" (*Weinberger* v *Romero-Barcelo* 1982).
- While the term "point source" is broadly defined, it does have certain limitations. In addition, EPA has exempted certain specific discharges (e.g.,

discharges in compliance with the instructions of an on-scene coordinator pursuant to the National Oil and Hazardous Substances Pollution Contingency Plan, storm-water runoff from crops or pastures, dredged or fill material) from NPDES permit requirements (EPA, § 122.3).

- The phrase "waters of the United States" includes almost all surface water. For example, many tributaries, impoundments, and wetlands are "waters of the United States" (EPA, § 122.2). On the other hand, waste-treatment systems, including certain treatment ponds or lagoons designed to meet the requirements of the CWA, are not "waters of the United States." Although groundwater is generally not considered to be "waters of the United States" (and is thus outside the scope of the NPDES program), discharges to groundwater may fall within the more expansive definitions of some states' regulatory programs.

EPA Permits

EPA procedures for issuing and modifying NPDES permits appear at 40 C.F.R. Part 124. To obtain an NPDES permit, a discharger must submit an application to the permitting authority, which is either the authorized state agency or EPA. A permit application requires detailed information relating to a facility and its discharges. Based on this information, the permitting authority will determine when and how discharges will be permitted.

If EPA is the permitting authority in a particular state, the permit application must be submitted to the appropriate EPA regional administrator at least 180 days before the date on which the proposed discharge is to begin or on which the current permit expires. After the permit application is submitted, the U.S. Army Corps of Engineers and other federal agencies (e.g., the Fish and Wildlife Service), as well as the state in which the discharge will occur, have an opportunity to comment on the application. If EPA concludes that the facility is a "new source," the applicant must also submit an environmental assessment (EA) as required by the National Environmental Policy Act (NEPA) (USC 4321).

After the permit application has been reviewed, EPA makes an initial determination as to whether the permit should be issued. If so, a draft permit, containing proposed effluent limitations, schedules of compliance, monitoring requirements, and so on (discussed below), is provided to the applicant, who may review and comment upon it. EPA is required to provide at least 30 days for the public to submit written comments and/or to request a public hearing, which must be held if there exists a significant degree of public interest in the proposed permit.

Following the public-comment period, EPA makes a final determination regarding the issuance of the permit. Within 30 days of EPA's final determination, any person may request an evidentiary hearing or legal review to reconsider the determination. The decision resulting from the evidentiary hearing may be appealed to the EPA Administrator.

A final permit is issued and becomes effective upon completion of the review proceedings. Compliance with the permit satisfies the requirements of the sections of the CWA on which the permit conditions are based, except for Section 307(a) of the CWA, which contains requirements pertaining to the discharge of toxic pollutants.

State Permits

State permitting procedures are generally similar to the federal permitting procedures, with certain exceptions. First, states are not required to provide an evidentiary hearing, although many do so. Second, provisions for judicial review of a permit issuance are those provided for in the state's administrative procedure act, rather than those provided in the federal Administrative Procedure Act (APA) (USC, § 551 et seq.). Finally, permit issuance by a state is not a federal action subject to NEPA, which otherwise would require that an environmental assessment (EA) and perhaps an environmental-impact statement (EIS) be completed.

Permit Contents

The principal purpose of an NPDES is to establish enforceable effluent limitations, which principally consist of numerical limitations on pollutant discharges. Effluent limitations, however, can also consist of restrictions based on visual observations (e.g., color), flow, temperature, pH, and other criteria.

Whether issued by a delegated state agency or EPA, NPDES permits contain standard boilerplate conditions that must be incorporated either expressly or by reference to the applicable state or federal regulations. Among boilerplate conditions are provisions addressing:

- the duty to comply with the permit
- the duty to take all reasonable steps to mitigate or prevent any discharges in violation of the permit that have a reasonable likelihood of adversely affecting human health or the environment
- the duty to provide information to the permitting authority
- the duty to monitor discharges and to submit discharge monitoring reports (DMRs)
- bypass and upset provisions

Notwithstanding the existence of these boilerplate conditions, permit provisions can and often should be negotiated.

Effluent-Treatment Criteria

The CWA employs two very different approaches in the management of direct-discharge-point sources. The predominant approach is known as the "technology-based" approach, which is based on national standards set by EPA. Under this approach, the type of pollution-control equipment a direct discharger must use is determined solely by evaluating the character of the pollutants discharged and the type of facility. It is irrelevant whether the pollutants are being discharged into a polluted lagoon or a pristine lake; under this approach, the type of pollutant coming out of the facility and the type of facility determine the required equipment.

The second approach is known as the "water quality–based" approach, in which states are largely involved. In contrast to the technology-based approach, this approach considers only the body of water and what is discharged into it. A state first designates a permissible use or uses for a water segment (e.g., drinking water, recreation, industrial discharge), and then determines the maximum amount of each pollutant that body of water can accept. This number effectively controls all those who discharge into that body of water.

Technology-Based Standards

Technology-based standards are the principal standards applicable to direct dischargers under the NPDES. These standards designate specific pollution-control equipment for designated industrial categories (EPA, § 405 et seq.). The type of pollution-control equipment required depends on two factors: (1) whether the pollutants in its waste stream are classified as "conventional," "toxic," or "nonconventional"; and (2) whether the point source in question is a "new source" or an "existing" source.

Type of Pollutant

EPA categorizes pollutants as follows:

> *Conventional Pollutants*—"Traditional" pollutants such as pH, total suspended solids (TSS), oil and grease, biochemical oxygen demand (BOD), and fecal coliform (EPA, § 401.16).
>
> *Toxic Pollutants*—particularly hazardous "priority" pollutants, including benzene, DDT, lead, mercury and asbestos (EPA, § 401.15).
>
> *Nonconventional Pollutants*—Also known as "gray-area" pollutants, this category includes all pollutants that are neither "toxic" nor "conventional,"

and includes common chemicals such as ammonia and chlorine (EPA, § 401.15).

Existing or New Source

Whether a facility is a new source or an existing source determines the type of technology-based treatment equipment a direct discharger must use. An existing source is any point source that was already in existence (and discharging pollutants) when EPA issued a new source performance standard (NSPS) for a substance in its pollutant stream (EPA, §§ 405-471). If a direct discharger begins construction of a facility and EPA has already proposed or finalized an NSPS for the pollutants that will be discharged from that facility, that facility is considered to be a "new" source. NSPSs are pollutant-specific pollution-control standards that specify required treatment protocols. There are different NSPSs for different kinds of industrial and agricultural operations ("source categories").

The type of treatment required is determined as follows:

Best Practicable Technology (BPT)—All *existing* point sources must utilize at least the "best practicable control technology currently available" (BPT) (USC, § 1314(b)(1)). BPT is based upon the national average effluent reduction achieved by current treatment systems and takes into consideration factors such as the age of the equipment, the feasibility of the control technology, and the process changes required and cost per unit reduction achieved.

Best Available Technology (BAT)—Most direct dischargers of "toxic" or "nonconventional" pollutants (including thermal pollution) must use the "best available technology economically achievable" (BAT) equipment to treat their waste streams (USC, § 1314(b)(2)). Under BAT, "economically achievable" is based on an industry-wide evaluation, and *not*, for example, on the financial situation of a particular direct discharger.

Best Conventional Technology (BCT)—All "conventional" pollutants must be treated with equipment considered "best conventional pollutant control technology" (BCT) (USC, § 1314(b)(4)). Like BAT, BCT takes into account the industry-wide economic feasibility of a particular control technology.

Best Available Demonstrated Control Technology (BADCT)—New sources must install BADCT for treatment of "conventional," "nonconventional," and "toxic" pollutants. BADCT serves as the basis for new source performance standards (NSPS), and is usually somewhat more stringent than BAT (USC, § 1316). It requires the maximum feasible reduction in pollutant discharge through the use of process changes, and in-plant and end-of-process controls. BADCT considers both the financial and environmental costs of a particular control technology.

Special Circumstances

EPA has also established special effluent limitation standards for certain situations that consider factors such as the specifics of a discharger's operational structure and local background water quality.

> *"Net/Gross" Credits*—Simply speaking, these credits ensure that direct dischargers that intake water simply for "throughput" purposes (i.e., heating or cooling processes) need not discharge water that is cleaner on the way out than it was on the way in (EPA, § 122.45(g)). A direct discharger will qualify for these credits if it can demonstrate that (1) its waste stream would meet its NPDES permit standards but for the "generic" pollutants already present in the intake water, and (2) the discharged generic pollutants are substantially similar to those in the intake water.
>
> *Variances*—Variances provide EPA and authorized state agencies with flexibility to modify national effluent standards when appropriate. There are a number of different types of variances, described below.
>> 1. *Fundamentally Different Factors (FDF) Variance*—By far the most common type of variance, an FDF variance may be granted where a literal interpretation of the regulations (i.e., "the letter of the law") would unfairly force a particular direct discharger to meet a stricter pollution treatment standard than the other members of its industry (EPA, § 403.13). Applications for FDFs are reviewed on a case-by-case basis, and are only granted if the direct discharger's facility is somehow "fundamentally different" from what EPA considered when developing that industry's treatment standards. Grounds for FDF variances include "non–water quality" environmental impacts caused by compliance with the treatment standard in question, or a severe disparity between the costs and benefits of the required treatment.
>> 2. *Water-Quality Variance*—The water-quality (WQ) variance is available only for "nonconventional" pollutants. To qualify, a direct discharger must show that the variance (1) will not require another point source to supplement its current treatment technology, (2) will not pose an "unacceptable risk" to human or environmental health, and (3) will, at the very least, comply with applicable BPT standards.
>> 3. *Economic Variance*—In contrast to the FDF variance, the CWA's economic variance is almost never granted. By law, it cannot be used to modify any BPT treatment standard, nor can it be used to modify treatment standards that apply to a "toxic" pollutant. The only time it can

be used is if (1) the variance will still "result in reasonable progress" toward pollution reduction, *and* (2) the proposed variance standard is the best "within the economic capability" of the direct discharger.

4. *Thermal Discharge Variance*—This variance is available to direct dischargers who can demonstrate that the thermal characteristics of their waste stream will not harm local aquatic or land species (EPA, §§ 125.70-125.73).

Water Quality–Based Controls

NPDES permits may contain not only technology-based controls, but where necessary to achieve water-quality goals, also may contain more stringent source-specific water quality–based controls. Unlike technology-based standards, however, water quality–based controls are not imposed on all dischargers. They are imposed on specific dischargers as an additional tier of protection when more stringent controls are needed to protect water quality.

Water quality–based controls are based on "water-quality standards." Water-quality standards are provisions of state or federal law that determine the use or uses of a particular body of water segment and, thus, the level of water-quality that will be necessary to support the designated uses. States are required to designate waterbody uses and establish appropriate water-quality standards for all bodies of water within the state, subject to review by EPA (USC, § 1313). States must also develop an "antidegradation policy," which is intended to maintain and protect existing water uses (EPA, § 131.12). Facilities that discharge into a waterbody that is not in compliance with that state's water-quality standards may be subject to whole effluent toxicity (WET) testing, which measures the aggregate toxic effect of the discharge on test samples (EPA, § 122.2).

Total Maximum Daily Loads

One specific manner in which water quality–based limitations are implemented is through the concept of the "total maximum daily load" (TMDL). A TMDL is the total amount of a pollutant that can be discharged into a particular waterbody by a discharger in a 24-hour period. TMDLs are specific numerical limits, and are incorporated directly into a discharger's NPDES permit.

Thermal Discharges

As noted above, the CWA specifically includes heat in its definition of the term "pollutant." However, if a source can demonstrate that a proposed thermal-discharge effluent limitation will require controls that are more stringent than necessary for the protection of a "balanced, indigenous population of shellfish, fish, and wildlife," the permitting authority may adjust these controls to a less-stringent level. What constitutes "balanced" and "indigenous" is often the subject of debate and is decided on a case-by-case basis.

Effluent Standards for Toxic Pollutants

Section 307(a)(1) of the CWA sets forth requirements for toxic pollutants. A toxic pollutant is a pollutant or a combination of pollutants that may cause death, disease, behavioral abnormalities, genetic mutations, or physical deformations in exposed organisms or their offspring. These pollutants are listed at 40 C.F.R. § 401.15; EPA may periodically add to or remove substances from the list. As noted above, discharge of any listed toxic pollutant is subject to the (BAT) standard designated by EPA for that category or class of point sources. Specific effluent standards for toxic pollutants appear at 40 C.F.R. Part 129.

Monitoring and Reporting Requirements

As discussed earlier, NPDES permits generally contain monitoring and reporting requirements. Dischargers must monitor their compliance with permit conditions on a regular basis and report these results to EPA on standard discharge monitoring report (DMR) forms. Monitoring is generally required to be performed at the point of discharge into the receiving water. Permits also typically specify the monitoring and analytical methods that must be used.

Monitoring information under the CWA must be kept for three years from the date of sample, measurement, report, or application. This includes:

- calibration and maintenance records
- original strip-chart recordings for continuous monitoring equipment
- copies of reports required by the permit
- records of all data used to complete the application for the permit

Permit Violations and Defenses

Technical violations of NPDES permit conditions, no matter how insignificant, are nonetheless violations. In other words, there is no *de minimis* defense to violations of NPDES permit conditions. However, narrow exceptions for certain types of excursions are typically incorporated directly into a permit. These are (1) "bypass" (i.e., intentional diversion of waste streams from any portion of a treatment facility), and (2) "upset" (i.e., unintentional, temporary noncompliance with technology-based permit effluent limitations) (EPA, §§ 122.41(m),(n)). A bypass is permitted only if necessary to prevent the loss of life, personal injury, or severe property damage, and if there is no feasible alternative. The upset defense requires proof that the upset was due to factors beyond the reasonable control of the discharger.

Permit Modification and Termination

EPA's permitting procedures (EPA, § 124) allow a permittee to transfer a permit easily to a new permittee to reflect a change in ownership. In addition, an NPDES permit can be "automatically" transferred to a new permittee if the current permittee notifies EPA or the state authority at least 30 days before the proposed transfer date, and provides the authority with other required information. EPA may revoke a permit for "cause."

6.3   INDIRECT DISCHARGERS

An "indirect discharger" is a discharger that does not discharge its waste stream directly into U.S. waters but discharges its pollutants into a publicly owned treatment works (POTW) by way of the local sewer system. Most facilities (including industrial, commercial, residential, and municipal facilities) are indirect dischargers.

Indirect dischargers are not subject to the NPDES permit program. They are, however, subject to pretreatment program requirements designed to protect the POTWs and waterbodies that lie downstream from them. POTWs are the familiar municipally run sewage-treatment plants charged with applying primary, secondary, and occasionally tertiary treatment to community waste streams.

POTWs have been designed to "recognize" and treat common residential and domestic waste-stream pollutants. They are not, however, designed to handle many industrial, commercial, institutional, and laboratory wastes. Many of these wastes can transit a POTW system completely "unrecognized" and untreated ("pass through"). Worse, if not pretreated by the indirect discharger, some of these wastes are toxic enough to interfere with ("Interference") or even completely disable a POTW's waste-treatment system and render its end sludge product unmarketable.

Pretreatment

EPA has promulgated national pretreatment standards that place restrictions on indirect dischargers (EPA, § 403 et seq.). Generally speaking, pretreatment standards are designed to ensure that effluent from direct and indirect dischargers is subject to similar levels of treatment. Separate treatment requirements exist for new and existing sources. Consequently, a particular indirect discharger may need to comply with as many as three sets of pretreatment standards:

National pretreatment standards—All indirect dischargers must comply with the general prohibition provisions of these standards, which prohibit the discharge of any pollutant that will either (1) pass through the POTW untreated or (2) cause interference with the POTW's biological treatment system. These standards also describe specifically prohibited discharges, including the discharge of pollutants that pose flammable or explosive hazards, have a pH of less than 5.0 or a temperature of greater than 104 degrees Fahrenheit, or consist of petroleum or mineral oils (EPA, § 403.5(b)).

Categorical pretreatment standards—These standards apply only to those indirect dischargers who are members of certain listed industrial categories (EPA, § 471). The nature of their waste streams require them to install specially designed treatment technologies.

Local pretreatment standards—POTWs with a total design flow of more than five million gallons per day are required to establish a local pretreatment program. Local pretreatment standards cover the vast majority of POTWs. These POTWs must establish pretreatment standards with which all local indirect dischargers must comply. These standards are based on an evaluation of the POTW's treatment capacity and the characteristics of its incoming waste stream. POTWs must ensure that the area's indirect dischargers are complying with these standards (USC, § 1319(f)).

An "affirmative defense" is available to indirect dischargers alleged to have violated the general prohibition against Pass Through and Interference standards (as well as certain specific prohibitions). These defenses are listed at 40 C.F.R. § 403.5(b)(3) – (b)(7). The defenses are available, if the discharger can demonstrate any of the following:

1. They did not know or have reason to know that their discharge would pass through or interfere with the POTW.
2. They were in compliance with the local pretreatment standard for the offending discharge at the time of the alleged violation.
3. If there is no such local pretreatment standard, the composition of their effluent did not change "substantially in nature or constituents" during the alleged violation (EPA, § 403.5(a)(2)).

Combined Waste Streams

If an indirect discharger mixes its pollutants with those of other dischargers, the resulting flow is known as a combined waste stream (CWS) (EPA, § 403.6). In this situation, EPA's "Combined Wastestream Formula" (CWF) determines each discharger's pretreatment obligations by calculating the pollutant concentration and flow rate of its waste stream and, thus, its contribution to the CWS. The CWF prevents dischargers from using dilution as a means to attain pretreatment standards (EPA, § 403.6(d)).

Removal Credits

Certain EPA-approved POTWs may grant "removal credits" to indirect dischargers for removal of "toxic" pollutants (EPA, § 403.7). Under this system, an indirect discharger need only pretreat its waste stream to the extent that after the waste stream subsequently undergoes treatment at the POTW, the final POTW effluent (and any sewage sludge that is generated) meets the applicable technology-based direct discharger standard. For most pollutants, this will be a BAT standard.

Monitoring and Reporting

Much like the discharge-monitoring report (DMR) requirements that apply to direct dischargers under the NPDES program, many indirect dischargers must monitor and record the flow rates and pollutant concentration of their waste streams (EPA, § 403.12). These records must be submitted to the local POTW or EPA-authorized authority on a regular basis.

Keeping Current

As mentioned earlier, there is no NPDES permit requirement for indirect dischargers, rather, pretreatment standards apply. In order to know which pretreatment technology must be employed, indirect dischargers must keep up to date with any changes in their local POTW pretreatment standards. With respect to national and categorical pretreatment standards, indirect dischargers are responsible for ensuring that any applicable modifications to these standards are incorporated into their program within 90 days. These standards typically are published in the *Federal Register* 90 days before they become effective. You can also contact your Regional EPA Office or EPA-authorized state environmental agency for current information and expected changes.

6.4　ENFORCEMENT/PENALTIES

Monitoring and Record Keeping

Section 308 of the CWA authorizes EPA to take almost any action reasonably necessary to gather information from point-source dischargers. This information can be used during the permit-development process or subsequently as part of the discharge-monitoring compliance program. All effluent data in these discharge records is available to the public. Dischargers may, however, petition EPA to withhold certain other information if it qualifies as "confidential business information" (EPA, § 122.7). EPA can also require effluent sampling, and the installation of discharge-monitoring equipment.

Entry and Inspection

EPA is also authorized to enter onto the premises of any facility where a point source is located (USC, § 1318(a)(4)(B)). Once on the premises, EPA may sample the waste stream, inspect treatment or other equipment associated with the point source, and inspect and copy pertinent records.

EPA may enforce the provisions of the CWA in a number of ways: by issuing compliance orders, assessing administrative penalties, initiating civil actions, or initiating criminal proceedings. In addition, citizens may bring suit against any discharger who is allegedly violating a standard or order authorized by the CWA.

Compliance Orders

Compliance orders are issued only after a discharger has failed to meet a final permit-compliance deadline. The order will specify a date by which the discharger must come into compliance with the terms of its NPDES permit (for direct dischargers) or its pretreatment requirements (for indirect dischargers). There is no fine or penalty involved unless the discharger still fails to come into compliance.

Administrative Penalties

EPA may also choose to assess administrative penalties for permit violations. There are two "tiers" of administrative noncompliance penalties available to the Agency:

- *Class I* penalties, which can be up to $10,000 per day of violation, up to a maximum of $25,000.
- *Class II* penalties, which can be up to $10,000 per day of violation, up to a maximum of $125,000.

Civil Actions

EPA can initiate a civil action against a CWA defendant in federal district court. Civil penalties can range up to $25,000 per day of violation, and are based on EPA's evaluation of factors such as the gravity of the offense, the defendant's past record/behavior, the defendant's current "attitude" or willingness to cooperate, and the economic benefit derived from noncompliance with the CWA.

Criminal Actions

Negligence

Under Section 309(c) of the CWA, EPA may criminally prosecute any person who "negligently" violates certain CWA provisions. Fines range up to $25,000 per day of violation and can be accompanied by up to one year of imprisonment.

Knowledge

EPA may criminally prosecute any person who "knowingly" violates certain provisions of the CWA. Fines range up to $50,000 per day of violation and can be accompanied by up to three years' imprisonment. Penalties for knowingly making false statements or tampering with the discharge-monitoring process are punishable by a fine of up to $10,000 and two years' imprisonment.

Endangerment

Under Section 309(c)(3)(A) of the CWA, EPA may criminally prosecute any person who "knowingly endangers" another person (i.e., places a person in "imminent danger of death or serious bodily injury"). Fines range up to $250,000 for individuals and $1 million for organizations. Individuals can also be sentenced to up to 15 years' imprisonment.

Blacklisting

Federal agencies are forbidden to contract with any persons convicted of the criminal actions above. For universities, schools, or hospitals having government contracts, this may be more than troublesome. EPA also maintains that it has the power to blacklist even without a conviction, even while the suit or administrative action is still pending (EPA, § 15).

Citizen Suits

Section 505 of the CWA allows any U.S. citizen to bring suit against a discharger who is allegedly violating a standard or order authorized by the CWA. It is important to note that the discharge-monitoring report (DMR) is a public document; it must be made available to anyone who requests it. It is often the basis for citizens' suits brought to enforce the CWA. The alleged violation(s) can be continuous or intermittent in nature, and the violation need not be ongoing. In order to bring suit, however, the citizen must have legal "standing," that is, she or he must have been directly injured in some way by the alleged violation. Injury can include the loss of recreational opportunities or damage to aesthetic values. Groups of citizens can also bring suit under the doctrine of "derivative standing"; if one of their members has been directly harmed by the alleged violation, the entire group may participate in the lawsuit. If successful, citizens can recover their costs and attorneys' fees, and the discharger will also be required to pay the appropriate civil penalties to EPA and/or the state.

Dischargers should make themselves aware of the restrictions as to when and how citizen suits can be brought. Citizens can only bring suit if:

1. The plaintiff is a U.S. "citizen" (USC, §§ 1365(g)).
2. The plaintiff has suffered a direct injury caused by the alleged violation.
3. The alleged violation occurred within the statutory period (i.e., within the "statute of limitations," which is generally five years (USC, § 2462), but which may vary between jurisdictions).
4. The violations cannot be "wholly past."
5. Neither EPA nor the EPA-authorized state agency are "diligently prosecuting" the alleged violation in an administrative or court proceeding.
6. The plaintiff has given 60 days' notice of intent to bring suit to both the discharger and EPA or the appropriate state agency.
7. If the U.S. is *not* a party to the suit, the plaintiff has served the Attorney General and the EPA Administrator with copies of any complaint.

Dischargers should also bear in mind that the CWA's citizen suit provisions can "work both ways"; universities, school systems, and hospitals may be able to bring suit as well. Thus, if the actions of another person violate the CWA or, conversely, the inaction of state or federal environmental authorities under the CWA is bringing harm to a facility's interests, the facility can seek to enforce the CWA itself by following the procedures outlined above.

6.5   STORM-WATER DISCHARGES

In 1987, Congress added Section 402(p) to the CWA, which was intended to establish a comprehensive framework for addressing storm-water discharges under the NPDES permit program. Under this provision, a schedule was established under which EPA was required to promulgate regulations and issue permits for various storm-water discharges. EPA was first required to regulate storm-water discharges "associated with industrial activity," and discharges from municipal non-combined storm sewers serving more than 100,000 persons.

In November 1990, EPA published a final rule, now codified at 40 C.F.R. § 122.26, which defined the term "storm-water discharge associated with industrial activity," and which established NPDES permit application requirements for such discharges from 11 major categories of industrial activities (Federal Register, November 16, 1990). While EPA interprets "storm water associated with industrial activity" as only that water directly related to manufacturing, processing, or raw-material storage areas at an industrial plant, according to EPA, this definition covers 100,000 facilities in the United States. All covered dischargers that discharge through a municipal separate storm sewer system or directly into U.S. waters must be covered by an NPDES permit. Storm-water discharges to sanitary sewer systems or POTWs are excluded.

In the November 1990 rule, industrial storm-water dischargers were presented with three permit application options: (1) submit an individual application, (2) become a participant in a group application, or (3) submit a Notice of Intent (NOI) to be covered under a general permit issued by EPA or an authorized state in accordance with the requirements of an issued general permit. In connection with the second option, in 1993 EPA published a proposed general permit for storm-water discharges associated with industrial activity (including discharges through certain municipal separate storm-sewer systems) for approximately 11,000 facilities in 29 industrial sectors in the states and territories without delegated NPDES authority. General permits for storm-water discharges associated with industrial activity and discharges associated with construction activity in states in which EPA is the permitting authority were published in 1992 (Federal Register September 9, 1992). Although these permits covered only 11 states and certain territories, many states with delegated NPDES authority have either developed their own general permits or have adopted EPA's general permits (Federal Register, November 19, 1993). Under the proposed permit, any facility falling within a covered sector would be allowed to apply for coverage under a five-year general permit. In addition to general conditions, each permit contains industry-specific provisions that outline storm-water pollution-prevention plan requirements, numeric effluent limitation requirements, and monitoring requirements.

As noted above, if a facility is not eligible for coverage under a general storm-water permit, it must submit a group application or an individual application. While universities, hospitals, and schools are not included *per se* within the 29 industrial sectors covered by the 1993 general permit, certain facilities (e.g., vehicle and equipment maintenance facilities, steam electric power–generating facilities) operated by universities, hospitals, and schools or by their contractors may be subject to storm-water requirements. Thus, storm-water permitting requirements for universities, hospitals, and schools must be addressed on a case-by-case basis. Fortunately, given its complexity, EPA has published a significant amount of information concerning storm-water requirements which should be consulted as storm-water issues arise (EPA 1992).

6.6   NONPOINT SOURCES

A nonpoint source (NPS) is any source of pollutants that is *not* a discernible, defined, discrete conveyance. NPSs include, for example, surface runoff from streets, automobile parking lots, construction sites, farms, ranches, and forestry and mining operations, and includes some point sources (e.g., irrigation return flows).

NPS pollution is addressed in Section 319 of the CWA, which encourages area-wide waste-treatment management by the states. Under Section 319, states must identify those waters that cannot attain or maintain adequate water quality due to NPS pollution, and must submit to EPA for approval state management programs for controlling NPS pollution. In addition to Section 319 requirements, EPA is authorized to promulgate regulations—known as best management practices (BMP)—governing a facility's site runoff, drainage, and waste disposal, which are typically NPSs (USC, § 1314(e)).

NPSs contribute the bulk of all the total suspended solids (TSS) and total dissolved solids (TDS) flowing into waterbodies in the United States. However, due in part to the vague and diffuse nature of this type of "source," at this time there are very few mandatory NPS requirements with which an operation or facility must comply. Thus, barring the existence of a specific local state program, for example, protection of groundwater/aquifers near "hot spots" (particularly polluted areas), regional estuary programs, and so on, there is no NPS treatment requirement or permit system affecting owners and operators of health and education facilities.

Those state nonpoint-source (NPS) programs that do exist are overwhelmingly voluntary and largely underfunded. Federal grant money has been made available to states that wish to implement an NPS program as part of their state clean-water strategy, but few have chosen to do so. States wishing access to these NPS funds must first conduct a State Assessment Report and develop a State

Management Program. Operators of health and education facilities interested in NPS issues should check with their state environmental agency for more information on its NPS policy and programs.

Owners and operators should also be aware of Section 306 of the Coastal Zone Management Act (CZMA) (USC, § 1455), which requires participating coastal states to implement management measures designed to reduce nonpoint pollution from agricultural, forestry, urban, marina, and hydromodification sources. These management measures will establish the pollutant-reduction practices and monitoring protocols required of affected facilities. State CZMA proposals were to be submitted to EPA and the National Oceanic and Atmospheric Administration (NOAA) in 1995; the management measures were scheduled to go into effect by January 1999.

6.7 WETLANDS

The construction or expansion of campus or hospital facilities may affect lakes, rivers, marshes, swamps, or other forms of government regulated "wetlands." The frustration for the construction planners is also manifest among the institution's environmental managers.

How best to define a "wetland" has been the subject of substantial debate in recent years (Haugrud 1993, Riverview 1985). The most common approach is based on an analysis of three factors:

1. *Hydrology*—The soil is saturated with water during the vegetation's growing season.
2. *Hydric Soils*—The upper portion of the soil is anaerobic due to water saturation for at least a portion of the growing season.
3. *Hydrophitic Vegetation*—Under normal conditions, at least 50% of the plants on the land in question are classified as "facultative" or "obligate" wetland plants.

Marshes, bogs, swamps, and freshwater ponds typically meet the definition of a "wetland." Wetland areas do not, however, need to be continually submerged or even damp. Although they can be permanent, wetlands may also be seasonal or even essentially random in character.

Oversight of both freshwater and saltwater wetlands is primarily the responsibility of the Army Corps of Engineers (the Corps), with oversight by EPA. Sections 301 and 502 of the CWA prohibit the discharge of "dredged" or "fill" material into U.S. surface waters without a Section 404 permit from the Corps. "Dredged material" is defined as all solid materials taken from U.S. waters. "Fill" is any material that has the *effect* of converting a wetland to dry land. These definitions can be very important in determining whether a discharger needs to obtain a Section 404 "fill" permit from the Corps or can proceed under a Section 402 NPDES "discharge" permit from EPA.

6.8   DREDGE AND FILL PERMITS

A Section 404 permit must be obtained from the Army Corps of Engineers before discharging dredged or fill material into U.S. surface waters. These permits become the focal point of the legal debate over a specific project (Percival et al. 1992). As discussed above, this broad category will include lakes, wetlands, and continuous and intermittent rivers and streams and their tributaries. A Section 404 permit is generally not required for discharges from the following activities: federally funded projects with an approved environmental impact statement (EIS); forestry, farming, and ranching operations; road construction; maintenance related to forestry or agriculture; and bridge, dam or levee operations (USC, § 1344(f)). Certain other activities, for example, survey activities, maintenance, temporary construction, and so on, may qualify for one of the CWA's "general permits" (EPA, § 330.6). If so, they need not be approved through the Section 404 process.

Removal or alteration of a wetland is often a politically and environmentally sensitive issue (Riverside 1985). Therefore, it is often prudent, from a financial and legal standpoint, to contact both EPA and the Corps as soon as the permit application process begins. One reason is that a variety of agencies and public-interest groups likely will be invited to comment on the permit application; it is best to be aware of any objections to the proposal as quickly as possible. Another reason is that EPA and the Corps have traditionally been at odds with regard to wetlands management. By establishing a working relationship with personnel at each agency early on, a dredge/fill discharge permit applicant can get a sense of how the application will be handled, who to contact when problems arise, and the likely outcome.

The steps permit applicants must follow in petitioning the Corps for a Section 404 permit are set forth at 33 C.F.R. § 325. However, permit applicants should not begin the process until after they have performed at least some form of preliminary engineering and environmental impact assessment on their own. This must be in writing, and should include:

1. the purpose of and reasons for taking the proposed action
2. alternatives to the proposed action (including "no action")
3. the costs and benefits of the proposed action—to water quality, flood control, vegetation, and wildlife (including any threatened or endangered species) and to cultural, community, and historical values

Permit applicants should also become familiar with 33 C.F.R. § 320, which describes the factors the Corps uses to evaluate permit applications.

6.9   SPILLS OF OIL AND HAZARDOUS SUBSTANCES

Under Section 311 of the CWA, EPA regulates the unintentional discharge of oil and other hazardous substances (EPA, §§ 116-117). EPA lists hazardous substances at 40 C.F.R. § 116.4. This list includes corresponding isomers and hydrates, as well as solutions and mixtures containing these substances.

The requirements of this provision are triggered only if a "reportable quantity" (RQ) of one of these substances is discharged. If a hazardous substance listed at 40 C.F.R. 116.4 is also listed in Table 302.4 (which appears at 40 C.F.R. Part 302), it is assigned the RQ for the substance appearing in Table 302.4. Otherwise, the RQ appearing at 40 C.F.R. § 117.3 applies. Discharges of oil are treated somewhat differently; generally, the RQ for oil is that amount that violates water-quality standards, causes a film or sheen on the water surface, or causes a sludge to be deposited either beneath the surface of the water or on shorelines (EPA, § 110.3).

When a person in charge of a vessel or an on-shore or offshore facility learns of a discharge of a hazardous substance to navigable waters or the adjoining shoreline in an amount greater than or equal to the RQ in a 24-hour period, that person must immediately notify the National Response Center (NRC) (800-424-8802) in accordance with Department of Transportation (DOT) requirements (EPA, § 153.203). Certain discharges authorized by the CWA or other authorities are exempt from notification requirements (EPA, §§ 117.11, 117.12).

EPA also requires owners of facilities that handle, transport, or store oil, and which, if "because of its location, could reasonably be expected to cause substantial harm to the environment by discharging into or on the navigable waters, adjoining shorelines, or the exclusive economic zone," to prepare and submit to EPA a written spill-prevention-control-and-countermeasure (SPCC) plan (EPA, § 112). However, these requirements generally only apply to facilities that handle substantial quantities of oil.

6.10   GRANTS AND RESEARCH PROGRAMS

Under Section 104 of the CWA, EPA is required to establish national programs for the prevention, reduction, and elimination of pollution. In implementing these programs, EPA is authorized to cooperate with educational institutions and to establish and maintain research fellowships at public, nonprofit private educational institutions or research organizations.

Under Section 105 of the CWA, EPA is authorized to make grants to any state, municipality, or intermunicipal or interstate agency to assist in the development of projects to (1) demonstrate a new or improved method of preventing, reducing, or eliminating discharges from sewers that carry storm water or (2) demonstrate advanced waste-treatment and water-purification methods, or new and improved methods of joint treatment systems. EPA may also provide grants to states or interstate agencies to demonstrate advanced treatment and environmental enhancement techniques in river basins or portions thereof.

EPA is also authorized to make grants and enter into contracts designed to carry out Section 301 of the CWA, which pertains to effluent limitations. Grants are also available to persons for research-and-development projects concerning pollution from agriculture and demonstration projects concerning sewage in rural and other areas in which conventional collection systems are not used.

6.11   WATER RIGHTS

Access to an adequate and dependable source of water is essential for any facility or operation. Although our modern infrastructure has effectively made this a moot issue for most of this century, there has been increasing concern and conflict over water rights—particularly in the American West and especially in this past decade (Haugrud 1993). Although American water law can vary widely from state to state, it can be divided into three main approaches: the "riparian" doctrine, the "prior appropriation" doctrine, and the "hybrid" approach, which combines the first two.

The *riparian doctrine* predominates in the eastern half of the country and is founded upon the concepts of ownership of property adjacent to the water source and a "reasonable" use of the resource. In contrast, the *prior appropriation doctrine*—which predominates in the western half of the country—is based on the principle of "first in time, first in right"; the first person to make beneficial use of the resource has laid claim to that portion of it. The *hybrid doctrine,* which incorporates aspects of both approaches, is increasingly popular and its principles are beginning to influence even those states that have traditionally been "pure" riparian (or prior appropriation) adherents.

Though somewhat obscure, these doctrines can have current, tangible effects. Facilities that hope to expand their physical size or their operational capacity may find their access to sufficient water in jeopardy or, occasionally, a facility simply hoping to maintain the status quo may discover a prior claim to part of its current supply. Because of the wide state-to-state variability in this area, affected facility owners and operators will need to check their state's water law to determine their facility's legal standing.

6.12   FUTURE DEVELOPMENTS

As this edition goes to press in 1996, Congress was considering reauthorization of the CWA. A strong deregulatory theme runs through current proposals and few, if any, new regulatory obligations are expected. It appears that the NPS program will remain voluntary and may even be weakened (states would not be required to draft an enforcement program). The storm-water-discharge permit system may also be changed into a voluntary system.

Wetlands may be classified into three categories based on their value relative to a national standard and their owners may become eligible for government compensation through a codification of the "regulatory takings" doctrine. Assuming these trends continue, the overall burden of water-related regulations on facility owners and operators may be modestly ameliorated over the next several years. However, it appears that the NPDES permit system for direct discharges and the pretreatment obligations of facilities sending wastewater to a POTW will remain largely unchanged.

REFERENCES

(CFR in text means Code of Federal Regulations)
33 Code of Federal Regulations § 153.203.
33 Code of Federal Regulations § 330.6.
40 Code of Federal Regulations § 15.
40 Code of Federal Regulations § 110.3.
40 Code of Federal Regulations § 112.
40 Code of Federal Regulations §§ 116–117.
40 Code of Federal Regulations §§ 117.11, 117.12.
40 Code of Federal Regulations § 122.2.
40 Code of Federal Regulations § 122.3.
40 Code of Federal Regulations §§ 122.41(m), (n).
40 Code of Federal Regulations § 122.45(g).
40 Code of Federal Regulations § 124.
40 Code of Federal Regulations §§ 125.70–125.73.
40 Code of Federal Regulations §§ 131.12.

40 Code of Federal Regulations § 401.15.
40 Code of Federal Regulations § 401.16.
40 Code of Federal Regulations § 403 *et seq.*
40 Code of Federal Regulations § 403.5(a)(2).
40 Code of Federal Regulations § 403.5(b).
40 Code of Federal Regulations § 403.6.
40 Code of Federal Regulations § 403.6(d).
40 Code of Federal Regulations § 403.7.
40 Code of Federal Regulations § 403.12.
40 Code of Federal Regulations § 403.13.
40 Code of Federal Regulations § 405 *et seq.*
40 Code of Federal Regulations §§ 405–471.
40 Code of Federal Regulations § 471.
EPA. 1993. Overview of the Storm Water Program. EPA-F-93-001. March 1993.
55 Federal Register 47990 (November 16, 1990).
57 Federal Register 41196 (September 9, 1992).
58 Federal Register 61146 (November 19, 1993).
Haugrud, K. Jack. 1993. Agriculture. In *Sustainable Environmental Law*, ed.
    Celia Campbell-Mohn, Barry Breen, and J. William Futrell, §8.2(A)(1). St.
    Paul: West Publishing.
Menell, Peter, and Richard Stewart. 1994. *Environmental Law and Policy.*
    Boston: Little Brown.
Percival, Robert, Alan Miller, Christopher Schroeder, and James Leape. 1992.
    Environmental Regulation: Law, Science and Policy.
5 United States Code § 551 *et seq.*
16 United States Code § 1455.
28 United States Code § 2462.
33 United States Code § 1251 *et seq.*
33 United States Code § 1313.
33 United States Code § 1314(b)(1).
33 United States Code § 1314(b)(2).
33 United States Code § 1314(b)(4).
33 United States Code § 1314(e).
33 United States Code § 1316.
33 United States Code § 1318(a)(4)(B).
33 United States Code § 1319(f).
33 United States Code § 1344(f).
33 United States Code § 1362(14).
33 United States Code §§ 1365(g).
42 United States Code § 4321.
United States v *Riverside Bayview Homes Inc.*, 474 U.S. 121 (1985).
Weinberger v *Romero-Barcelo*, 456 U.S. 305 (1982).

CHAPTER 7

Transportation of Hazardous Materials

7.1 OVERVIEW OF RISKS FROM TRANSPORTATION

Why read this chapter? Your job description probably does not include "hazardous waste compliance" expressly, and most educational and health-care institutions are not transporters of hazardous material. The typical institution handles some modest amount of these specially regulated wastes, and it routinely hires contractors who prepare the necessary shipping papers and then pack, label, and transport the material.

There are two important reasons for the safety professional to understand transportation of hazardous materials. First, the regulations impose obligations on shippers, that is, any educational or health-care institution that performs any of the following activities: (1) generates waste and arranges for its transport (2) returns hazardous materials to their supplier, or (3) tenders empty packaging containing the residue of hazardous materials. The shipper retains these responsibilities even if the transporter promises to fulfill the shipper's responsibilities. So this type of obligation has a long "tail."

Second, as the earlier chapter on waste management discussed, waste generators face perpetual liability for any harm caused by their waste. Thus, an important part of any waste-management program is careful selection of a responsible transporter and disposal facility. Your decisions on the selection of waste vendors are potentially subject to expensive future consequences if you don't understand the legal consequences of your shipment practices.

How does a regulatory manager assess potential transporters? To properly evaluate a transporter, a manager must have a clear picture of the transporter's regulatory obligations in addition to applying normal business cost, reliability, and reputation criteria. This chapter provides an overview of the U.S. Department of Transportation's (DOT) regulations, which establish a comprehensive classification, hazard communication, packaging, labeling, and training plan designed to ensure that hazardous materials are safely transported.

7.2   WHO IS COVERED BY THE REGULATIONS?

Scope

At present, hazardous-materials regulations of the Department of Transportation (DOT) apply to: (a) all interstate shipments of hazardous materials and (b) intrastate (within one state) shipments by motor vehicle of hazardous wastes, hazardous substances, flammable cryogenic liquids, and marine pollutants. DOT proposed in 1993 to extend the federal hazardous-materials regulations to all intrastate shipments of hazardous materials (DOT 1993).

These regulations apply to shippers, truck drivers and other transport personnel, and packaging manufacturers and reconditioners involved in any aspect of the transport of hazardous materials. DOT's hazardous materials regulations govern the domestic shipment of hazardous materials by road, rail, air, and vessel. Some domestic air carriers and all international air shipments are governed by the International Civil Aviation Organization's (ICAO) Technical Instructions.[1] Shipments by ocean vessel are subject to the International Maritime Dangerous Goods (IMDG) Code.

Activities Covered

The activities addressed by DOT's hazardous-materials rules include:

> preparation of material for transport
> loading of materials
> domestic movement of materials by any mode
> unloading of materials
> storage incidental to transport

Storage incidental to transport would include temporary storage while the material is in transit, or on its way to its final destination (USC, § 1802(15)).

---

[1] Some air carriers may follow the International Air Transport Association's (IATA) Technical Instructions. These are based on the ICAO Technical Instructions, although in some instances they are stricter. DOT officially recognizes only the ICAO Technical Instructions.

Shippers

Under DOT's regulations, a shipper is any entity or person placing hazardous materials into commerce. This definition is broad in scope and includes a university shipping infectious materials or hazardous wastes via a third-company motor carrier. However, DOT has issued an interpretation indicating that universities transporting hazardous materials in university vehicles for noncommercial purposes are not subject to DOT's hazardous-materials regulations. The term "shipper" also applies to anyone returning hazardous materials to its supplier, or shippers tendering empty packaging containing the residue of hazardous materials.

A shipper of hazardous materials must fulfill a number of duties under the regulations, which are broadly defined as follows:

- classify its hazardous materials tendered for shipment
- complete the shipping documentation for the material
- mark and label the package, and placard the vehicle, if required, for the hazards associated with the material
- ensure that the packaging utilized to transport the hazardous materials is compatible with the material and meets the design specifications and performance tests detailed in DOT's regulations

Truck Drivers/Transport Personnel

Truck drivers and other transport personnel involved in the shipment of hazardous materials are subject to DOT's regulations unless otherwise exempted. A driver is responsible for the safe loading/unloading of the hazardous material, to the extent he performs those functions, and for ensuring the safe movement of the hazardous-materials shipment from origin to destination point. As most materials shipped from universities or hospitals likely will move by motor carrier, this chapter will focus primarily on motor-carrier requirements (DOT, Part 177). There also are requirements specific to rail transportation (49 C.F.R. Part 174), air (49 C.F.R. Part 175), and vessel (49 C.F.R. Part 176). Drivers of motor vehicles hauling hazardous materials are required to obtain a commercial drivers license, and, in most instances, a hazardous-materials endorsement (DOT § 383).

Manufacturers and Reconditioners of Packaging

Manufacturers and reconditioners of hazardous materials packaging are responsible for constructing the packaging in conformance with the packaging specifications, ensuring that the packaging satisfies the required performance tests and marking the packaging with the United Nations Certification marks.

7.3   CLASSIFYING HAZARDOUS MATERIALS FOR TRANSPORT

The hazardous materials regulations are triggered by the offering for transport of any material that satisfies at least one of the definitions of a hazardous material. DOT regulates materials as hazardous because they are either known to be hazardous or because the material's hazard criteria satisfies DOT's definition of a hazardous material.

There are nine different hazard classes under DOT's regulatory scheme. Each hazard class has its own definition section which establishes the hazard criteria for that class. These definitions are complex and have a number of exemptions, so reference to the appropriate regulatory section is essential when classifying a material (see Table 7–1). Some of these classes are divided into hazard divisions because of their unique concerns or properties. Generally the hazard classes/divisions and their relevant section numbers are as follows:

Explosives (Class 1), Section 173.50
A material that functions by exploding or because a chemical reaction may function in a similar manner.
Gases (Class 2), Section 173.115
Class 2 includes the following divisions:
- Division 2.1—Flammable gas. A gas that is flammable or ignitable at a certain pressure when in a specific concentration range of air.
- Division 2.2—Nonflammable, nonpoisonous compressed gas and asphyxiant gases; a gas that does not meet the definition of a Division 2.1 or 2.3 gas.
- Division 2.3—Gas poisonous by inhalation. A gas that is either known to be toxic to man or is presumed so because of animal test data.
Flammable liquids (Class 3), Section 173.120
A liquid that has a flashpoint of not more than 141°F, or any material with a flashpoint at or above 100°F that is intentionally heated and offered for transport at or above its flash point in a bulk packaging and does not meet one of the listed exceptions.
Flammable solids (Class 4), Section 173.124
- Division 4.1—Flammable solid includes wetted explosives, self-reactive materials, and readily combustible materials.

- Division 4.2—Spontaneously combustible material, includes materials such as pyrophoric materials and self-heating-materials.
- Division 4.3—Dangerous-when-wet material includes a material that when wetted is likely to ignite spontaneously and give off flammable or toxic gas.

Oxidizers/Organic Peroxides (Class 5)

- Division 5.1—Oxidizers, Section 173.127. A material that yields oxygen and causes or enhances combustion of other materials.
- Division 5.2—Organic peroxides, Section 173.128. An organic compound containing oxygen in the bivalent -0-0- structure which is a derivative of hydrogen peroxide and where one hydrogen atom has been replaced by organic radicals.

Poisons/Infectious Substances (Class 6)

- Division 6.1—Poisons, Section 173.132. A material that is known to be toxic to man or because of oral, dermal, or inhalation animal data is presumed toxic.
- Division 6.2—Infectious substances, Section 173.134. A material that contains or is contaminated with an infectious substance. This definition includes regulated medical waste (DOT 1994)

Radioactives (Class 7), Section 173.403

A material that is radioactive.

Corrosives (Class 8), Section 173.136

A material that causes irreversible damage to the skin or has a severe corrosion rate on steel or aluminum.

Miscellaneous Hazard (Class 9), Section 173.140

The miscellaneous hazard class includes materials that do not satisfy any other hazard criteria but fall into the following categories:

- Hazardous wastes regulated by the U.S. Environmental Protection Agency
- EPA-designated hazardous substances and their mixtures that are listed in Appendix A to Section 172.101 and are shipped in one package in a quantity exceeding their reportable quantity
- Marine pollutants listed in Appendix B to Section 172.101, or their mixture or solutions, when packaged in a concentration which equals or exceeds:
  (a) Ten % by weight of the solution or mixture for materials listed in the Appendix or
  (b) One % by weight of the solution or mixture for materials that are identified as severe marine pollutants (SMP)

In addition to the hazard class designation, most hazard classifications (with the exception of Classes 1, and 2 and Division 6.2) are divided into three pack-

Table 7-1.   Checklist for Hazardous-Materials Shipments

A. Classification
Determine whether the material is specifically listed in the Hazardous
Materials Table.   _____

If the material is not specifically listed, determine whether the material
meet any of the nine hazard definitions.   _____

Select the proper shipping name for the material.   _____

Determine the appropriate hazard class or classes.   _____

Select the proper identification numbers.   _____

Select the proper packing group.   _____

B. Packaging/Placarding
Determine the mode(s) of transport to ultimate destination.   _____

Determine and select the packaging according to Column 8 of the
Hazardous Materials Table.   _____

Determine whether an exception is available.   _____

Determine whether any special provisions apply.   _____

Ensure package is properly filled and that no material remains on
outside of container.   _____

Ensure inner packaging is cushioned as required and that packagings
are positioned as required (especially for liquids).   _____

Mark the packaging with all required marks.   _____

Select the proper labels and apply as required.   _____

Select the proper placard(s) and apply if required.   _____

C. Shipping Papers/Emergency Response
Prepare the shipping papers with the required hazard information.   _____

Certify the shipment.   _____

Indicate proper placards on shipping paper, if not attached.   _____

Supply emergency-response information.   _____

D. Loading/Bracing/Securing Shipment
Ensure the package is properly loaded to prevent damage to container.   _____

Determine whether any packages need to be segregated from other types
of hazardous materials.   _____

Ensure the package is secured and brace as required to prevent movement
during shipment.   _____

ing groups, with Packing Group I indicating high hazard materials; II indicating moderate hazard; and III low hazard. The regulations specify the criteria for determining the correct packing group for a given hazardous material.

7.4   DETERMINING HAZARD STATUS OF YOUR SHIPMENT

The institutional safety manager is expected to differentiate between the various regulatory definitions and "be right the first time." As a shipper, you are responsible for classifying your hazardous material for shipment. If you determine that the material is hazardous, then the shipment will be regulated by the Department of Transportation. Be aware, however, that somewhat different hazard definitions have been developed by DOT, OSHA, and EPA and, therefore, a material that is defined as hazardous by EPA or OSHA is not necessarily hazardous for transportation purposes.

There are a number of steps that must be followed to determine whether a material is regulated by the DOT. This process may be simplified if the material was supplied by another company and that company has provided a material-safety data sheet (MSDS) as required by OSHA (OSHA 1910.1200). Generally, suppliers list all relevant hazard-communication information on the MSDS, including all relevant DOT-required information for the material. Therefore, if the shipper has the supplier's MSDS, and there have been no significant changes to the material by the shipper, then the shipper should be able to use a reliable supplier's MSDS. However, reliance on the supplier's MSDS does not relieve the shipper of the duty to classify and package the material properly, and mark, label, and placard for associated hazards.

Since all shippers are responsible for understanding the classification scheme, Figure 7.1 and the following description review the steps that must be taken to properly classify your hazardous material.

Consult the Hazardous-Materials Table

The Hazardous-Materials Table which appears at 49 C.F.R. § 172.101 specifically lists hundreds of materials that have been deemed to be hazardous when in transport. Each material is listed by its proper shipping name in Column 2. Once the proper shipping name is located, the shipper can determine the material's hazard class (Column 3); United Nations' worldwide standardized chemical identification number, or "UN number" (Column 4); packing group (Column 5); the pre-

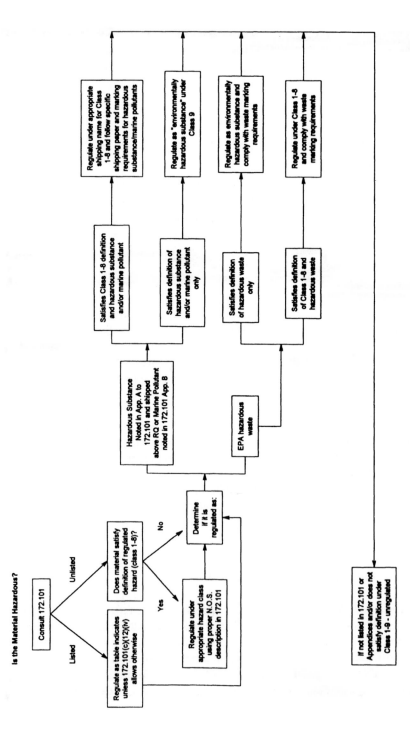

Figure 7.1. Classifying materials under DOT's regulations.

scribed labeling (Column 6); any special provisions relating to the transport of the material (Column 7); and the appropriate packaging (Column 8).

Compare the Material's Characteristics with the Hazard Criteria

Even if a material is not specifically listed, it may still be a hazardous material in the event that it meets at least one of the definitions of a DOT-regulated hazardous material. Therefore, a shipper must be familiar with the material's hazard characteristics (*e.g.*, flash point, toxicity, corrosivity, etc.).

The shipper should review the first eight hazard definitions referenced in Section 7.03 and compare the material's hazard characteristics with the regulatory definitions. If a shipper determines that the material satisfies one of the hazard criteria, then he must use the correct generic shipping name that describes that material's hazard. For instance, a material that satisfies the definition of a poisonous liquid but is not specifically listed in the table would have the proper shipper name "Poisonous liquids, n.o.s" (Division 6.1).

Determine If the Material Satisfies the Definition of Class 9

Regardless of whether the material is already regulated as a hazardous material under Hazard Classes 1–8, the shipper also must determine whether the material is a regulated hazardous substance, marine pollutant, or an EPA-designated hazardous waste.

If a material satisfies at least one of the definitions of a hazardous material under Classes 1–8, and also satisfies the definition of a Class-9 material because it is either a hazardous substance, marine pollutant, or an EPA-designated hazardous waste, the shipper must classify the material under the appropriate primary hazard class of 1–8. The shipper also must indicate on the shipping paper that the material is a hazardous waste, hazardous substance, or marine pollutant and comply with the appropriate labeling and marking requirements.

If a material does not satisfy any of the definitions in Classes 1–8, but does satisfy the definition of either a hazardous substance, marine pollutant, or an EPA-designated hazardous waste, then it is regulated as a Class-9 hazardous material. Again, the shipper must comply with the shipping paper requirements and the appropriate labeling and marking requirements.

7.5   SPECIAL CONSIDERATIONS IN CLASSIFICATIONS

Classifying Mixtures

If a material is a mixture of a hazardous material that is listed in the Hazardous Materials Table, then the shipper must determine how to classify the material (DOT, § 172.101(c)(10)(i)). Generally, a mixture that consists of a listed hazardous material and nonhazardous material is classified according to the proper shipping name, hazard class, and packing group of the hazardous-material ingredient, unless one of the following conditions exists:

- The physical state of the mixture is different from the pure material so that the authorized packaging is inappropriate for the mixture.
- The shipping description indicates that the proper shipping name applies only to pure or technically pure material.
- The hazard class, packing group, or subsidiary hazard of the mixture is different from that for the listed material.
- The listed material is a material poisonous by inhalation, but the mixture no longer meets the criteria of a poison by inhalation.
- There is a significant change in the measures to be taken in emergencies for the mixture.
- The material can be appropriately named with a proper shipping name that describes its intended application.

If at least one of these conditions is triggered, then the shipper must determine which generic shipping name is appropriate to the material. In making this determination, the shipper must look at the hazards associated with the material and the physical state of the material.

Subsidiary Hazards

A material may have more than one hazard if it meets the criteria of more than one hazard class. Materials with more than one hazard are said to have "primary" hazards and "subsidiary" hazards. The primary hazard is the hazard that takes precedence over all other hazards in terms of where it appears on the shipping documentation and labeling/placarding.

Many materials specifically listed in the Hazardous Materials Table have subsidiary hazards assigned to them that appear after the primary hazard. If a material is being shipped under a generic classification (not otherwise specified or n.o.s.) and the material meets the definition of more than one hazard class, then the shipper must classify the material according to the highest applicable hazard class of the following hazard classes:

1. radioactives (Class 7)
2. poisonous gases (Division 2.3)
3. flammable gases (Division 2.1)
4. nonflammable gases (Division 2.2)
5. poison (Division 6.1) (Packing Group I materials poisonous by inhalation only)
6. pyrophoric liquids (Division 4.2)
7. self-reactive (Division 4.1)
8. flammable liquids, corrosives, flammable solids, spontaneously combustible, dangerous when wet, oxidizers, poisonous liquids, or solids other than Packing Group I PIH
9. combustible liquids
10. miscellaneous hazardous materials

To determine the order or precedence for the materials grouped under Number 8, it is necessary to refer to the Precedence of Hazard Table (DOT § 173.2a(b)); see Table 7–2.

7.6   HAZARD COMMUNICATION
FOR TRANSPORTED MATERIALS

The near-universal coverage of the Occupational Safety and Health Administration's Hazard Communication Standard (OSHA 1910.1200) is augmented, for transported chemicals, by rules of the Department of Transportation (DOT) relating the same subject. DOT seeks to educate and induce caution among drivers, loaders, and other engaged in the transporting of hazardous materials.

Labeling, Marking, and Placarding

As a shipper of hazardous materials subject to DOT's requirements, educational or health-care institutions must ensure that each package containing a DOT-regulated hazardous material is properly labeled and marked (DOT, § 173.22). Labeling clearly identifies the hazards posed by the particular material. There are designated labels for each hazard class and division. Materials must be labeled for their pri-

Table 7–2. Precedence of Hazard Table/DOT (Hazard class and packing group)

	4.2	4.3	5.1 I[1]	5.1 II[1]	5.1 III[1]	6.1,I dermal	6.1,I oral	6.1 II	6.1 III	8,I liquid	8,I solid	8,II liquid	8,II solid	8,III liquid	8,III solid
3 I						3	3	3	3	3	(3)	3	(3)	3	(3)
3 II						3	3	3	3	8	(3)	3	(3)	3	(3)
3 III						6.1	6.1	6.1	3[4]	8	(3)	8	(3)	3	(3)
4.1 II[2]	4.2	4.3	5.1	4.1	4.1	6.1	6.1	4.1	4.1	(3)	8	(3)	4.1	(3)	4.1
4.1 III[2]	4.2	4.3	5.1	4.1	4.1	6.1	6.1	6.1	4.1	(3)	8	(3)	8	(3)	4.1
4.2 II		4.3	5.1	4.2	4.2	6.1	6.1	4.2	4.2	(3)	8	(3)	4.2	(3)	4.2
4.2 III		4.3	5.1	5.1	4.2	6.1	6.1	6.1	4.2	(3)	8	(3)	8	(3)	4.2
4.3 I			5.1	4.3	4.3	6.1	4.3	4.3	4.3	4.3	4.3	4.3	4.3	4.3	4.3
4.3 II			5.1	4.3	4.3	6.1	4.3	4.3	4.3	8	8	8	4.3	4.3	4.3
4.3 III			5.1	5.1	4.3	6.1	6.1	6.1	4.3	8	8	8	8	4.3	4.3
5.1 I[1]						5.1	5.1	5.1	5.1	5.1	5.1	5.1	5.1	5.1	5.1
5.1 II[1]						6.1	5.1	5.1	5.1	8	8	8	5.1	5.1	5.1
5.1 III[1]						6.1	6.1	6.1	5.1	8	8	8	8	5.1	5.1
6.1 I, Dermal										8	6.1	6.1	6.1	6.1	6.1
6.1 I, Oral										8	6.1	6.1	6.1	6.1	6.1
6.1 II, Inhalation										8	6.1	6.1	6.1	6.1	6.1
6.1 II, Dermal										8	6.1	8	6.1	6.1	6.1
6.1 II, Oral										8	8	8	6.1	6.1	6.1
6.1 III										8	8	8	8	8	8

[1] There are at present no established criteria for determining Packing Groups for liquids in Division 5.1. For the time being, the degree of hazard is to be assessed by analogy with listed substances, allocating the substances to Packing Group I, great; II, medium; or III, minor danger.

[2] Substances of Division 4.1 other than self-reactive substances.

[3] Denotes an impossible combination.

[4] For pesticides only, where a material has the hazards of Class 3, Packing Group III, and Division 6.1, Packing Group III, the primary hazard is Division 6.1, Packing Group III.

mary hazard and for subsidiary hazards if specifically required by the regulations, or if otherwise required (DOT, § 172.402). Labels must meet the size specifications in the regulations. There are special provisions in DOT's regulations for the labeling of mixed and consolidated packaging and for subsidiary hazards.

There are two types of marking requirements under the regulations—hazard-communication markings and package-manufacture markings. This part focuses in hazard-communication markings. All packages, freight containers, and vehicles must be properly marked with hazard-communication information (DOT, §§ 172.300–.338). There are specific regulations governing the size, color, and placement of markings. Nonbulk packages generally must be marked with the following:

1. the proper UN number for a material
2. the material's proper shipping name
3. the name of the consignor or consignee (DOT, § 172.301)

Bulk packaging must be marked with the UN number (DOT, §172.302).
There are special marking requirements for the following types of materials:

- materials shipped in exemption packaging (DOT, §§ 172.301 and .302)
- radioactive materials (DOT, § 172.310)
- liquid hazardous materials in nonbulk packaging (orientation arrow marking) (DOT, § 172.312)
- poisonous hazardous materials (DOT, § 172.313)
- packaging containing materials classified as ORM-D (DOT, § 172.316)
- explosives (DOT, § 172.320)
- marine pollutants (DOT, § 172.322)
- hazardous substances in nonbulk packages (DOT, §172.324)
- elevated temperature materials (DOT, § 172.325)
- materials shipped in portable tanks, cargo tanks, tank cars and multiunit tank car tanks and other bulk packaging (DOT, §§ 172.326–.331)

In addition to labeling/marking requirements, all motor vehicles or freight cars/tank cars must be placarded to indicate the hazard posed by the particular material. Placards are similar in design to labels, although they are larger and must meet certain durability and visibility tests.[2] Shippers are responsible for providing the placards to the driver when offering a hazardous material for transport by motor carrier (DOT, § 172.506). When offering a material for transport by rail, shippers are responsible for affixing the placard to the railcar, or where the railcar is not in their possession, indicating on the shipping paper the required placards (DOT, §172.508).

---

[2] The regulations governing placarding are located at DOT, §§ 172.500–.560.

Shipping Paper Requirements

All bills of lading, hazardous-waste manifests, and other shipping papers must be completed by the shipper (DOT, §§ 172.200–.205). Each hazardous material must be separately listed on the shipping paper, and either be entered first, entered in a contrasting color, or identified by the entry of an "X" placed before the proper shipping name in a column captioned "HM." The shipping paper must include the following:

- the proper shipping name as listed in Column 2 of the Hazardous Materials Table
- the hazard class or division for that material
- the UN identification number
- the packing group
- the total quantity of material (DOT, §§ 172.202)

Additional descriptions are required on the shipping paper for:

- materials moving under an exemption
- shipments of hazardous substances
- materials shipped under a limited quantity exemption
- shipments of empty packaging containing a residue of a hazardous material
- materials shipped under an n.o.s. or other generic description
- materials designated marine pollutants (DOT, § 172.203)

Shippers must certify on the shipping paper that the hazardous materials have been properly classified, labeled, marked, placarded, and packaged using the following language:

This is to certify that the above-named materials are properly classified, described, packaged, marked, and labeled and are in proper condition for transportation according to the applicable regulations of the Department of Transportation (DOT, § 172.204(a)(1)).

The shipping paper also must include a 24-hour emergency-contact phone number that is clearly indicated. This number may be the shipper's own established emergency hot line or a paid service such as the Chemical Manufacturers Association's CHEMTREC toll-free number. However, the person answering the hot line must be knowledgeable and able to direct emergency personnel immediately as to the hazards posed by the material.

Other emergency response information must be included on or with the shipping paper indicating:

the proper shipping name of the material
immediate health hazards
risks of fire or explosion
immediate precautions to be taken in the event of an accident
initial methods for handling spills or leaks in absence of fire
preliminary first-aid measures (DOT, § 172.602)

This information should be written so that it is easily understood by emergency-response personnel. Carriers must maintain the information with the shipping paper in the cab and have it accessible at all times. Facility operators at each location where a hazardous material is received, stored, or handled during transport must maintain the information in a location where the hazardous material is present and keep it immediately accessible in the event of an incident.

7.7   Properly Packaging Hazardous Materials

All hazardous materials must be packaged in containers authorized for that material in Column 8 of the DOT regulations' Hazardous Materials Table. The Hazardous Materials Table shows three possible packaging references: packaging exceptions available (Column 8a); nonbulk packaging (Column 8b); and bulk packaging (Column 8c). Under each column, a section of the regulations is identified. This section is known as a "packaging authorization." To fully explain the different subcolumns under Column 8, this section will focus first on nonbulk and bulk packaging authorizations and finally on the packaging exceptions.
    Nonbulk packaging is defined as:

Packaging that has a maximum capacity less than 450 liters (119 gallons) for a liquid; a maximum net mass less than 400 kg (882 pounds) and a maximum capacity less than 450 liters for a solid; or a water capacity greater than 454 kg (1,000 pounds) or less as a receptacle for a gas (DOT, § 171.8).

Bulk packaging is defined as:

Packaging other than a vessel or a barge, including a transport vehicle or freight container, in which hazardous materials are loaded with no intermediate form of containment, and which has a maximum capacity greater than

450 liters (119 gallons) as a receptacle for a liquid; a maximum net mass greater than 400 kg (882 pounds) and a maximum capacity greater than 450 liters as a receptacle for a solid; or a water capacity greater than 454 kg (1,000 pounds) as a receptacle for a gas.

Bulk packaging is further divided into a subgroup of packaging known as an intermediate bulk container, which is defined as:

A rigid or flexible, portable packaging, other than a cylinder or portable trunk, which is designed for mechanical handling (DOT, § 171.8).

Each packaging-authorization section references design types allowed for that material. Design specifications are provided for each packaging type (e.g., fiberboard box, plastic drums, steel drums, etc.) (DOT, Part 178). Additionally, nonbulk packaging and intermediate bulk containers must meet specified performance tests (e.g., drop, stacking, vibration, leakproofness, hydrostatic). The types of tests required vary depending on the type of packaging (e.g., drum, fiberboard box, etc.), the physical state of the material (liquid, solid, gas), and the packing group level (I, II, or III). Bulk packaging (e.g., cargo tanks, portable tanks) has its own design and test requirements (DOT, Part 179).

It is the shipper's responsibility to ensure that the proper packaging is used in the shipment of a hazardous material. Shippers, however, likely do not have the facilities for testing the packaging, therefore, they must rely on the packaging supplier. Consequently, shippers should receive certification from their packaging suppliers that the packaging satisfies the necessary design specifications and performance tests. One method of identifying hazardous-materials packaging is by using the markings placed on the package by the manufacturer. All hazardous-materials packaging must be marked with the following:

- UN symbol
- information indicating the packaging's construction material and design type
- letters indicating packing group
- designation of specific gravity or mass for which packing has been tested
- result of the hydrostatic pressure test (liquids only)
- year of manufacture
- country of manufacture
- name and address of the manufacturer
- minimum thickness (metal or plastic drums or jerricans only) (DOT, § 178.503)

7.8   USE OF EXCEPTIONS IN TRANSPORTATION

As with any regulatory progress, there are a number of exceptions available under the regulations (DOT, Part 173). Some are very specific (such as agricultural operations) and others are much broader. This section will briefly review the primary exceptions available.

Exceptions for Small Quantities

Small quantities of the following materials are exempt from DOT's hazardous-materials regulations under certain conditions:

- Class 3 (flammable liquids)
- Division 4.1 (flammable solids)
- Division 5.1 (oxidizers)
- Division 5.2 (organic peroxides)
- Division 6.1 (poisons)
- Class 8 (corrosives)
- Class 9 (miscellaneous)
- Class 7 (radioactive materials that also meet the definition of one of the above materials)

Materials shipped under the small-quantity exception must comply with the following:

- For all materials except Division 6.1 (poisons) in Packing Group I, the maximum quantity of material per inner receptacle cannot exceed 1 ounce (30 ml).
- Liquids and solids in Division 6.1 (poison) Packing Group I may not exceed 0.04 ounces (1 ml).
- Radioactive materials must not exceed certain activity levels;
- All materials must be packaged and tested according to the special provisions for small quantities (DOT, § 173.4(a)(2)–(11)). Unlike other exemptions, the small-quantity exemption has its own drop-test requirements for packagings.

Limited-Quantity Exceptions

The limited-quantity exception is different from the small-quantity exception in that it generally provides exceptions to the testing requirements for packagings, with the exception of the vibration test. For a material to be eligible for this exception, Column 8A of the Hazardous Materials Table must reference the relevant exception in the listing for the material. If a particular listed material does not reference a section in Column 8A, or the word "none" is printed in the column, then there is no limited-quantity exception available for that material.

If Column 8A references a section of the regulations, then the shipper must review that section to determine how to comply with the exception. While the specifics of the limited-quantity exceptions vary depending on the hazard class and packing group of the material, there are some provisions generic to the exception. First, all materials shipped under a limited-quantity provision must be packaged in combination packages (an inner and outer packaging). The section reference for the exception references the specific packaging types available for that material. Each inner container has a weight limit that varies depending on the hazard class and packing group of the material. The gross weight of the container plus contents may not exceed 66 pounds.

Exceptions for Shipment of Waste Materials

There are several exceptions for waste materials that may be relevant at a given time.

Closed-Head Drums

A material that is a hazardous waste and is required to be shipped in a closed-head drum (i.e., a drum with a 7.0 cm or less bung opening), and the hazardous waste contains solids or semisolids that make its placement in a closed-head drum impracticable, may be shipped in an equivalent open-head drum (DOT § 173.12(a).

Lab Packs

Certain waste materials are excepted from the packaging requirements of DOT's regulations if packaged in combination packages that satisfy the performance requirements outlined in the exception and are transported for disposal or recovery by motor carrier (DOT, § 173.12(b)). The following classes or divisions of materials are covered by the lab-pack exception:

- Hazard Class 3 (flammable liquids)
- Division 4.1 (flammable solids)
- Division 4.2 (spontaneously combustible)
- Division 4.3 (dangerous when wet)
- Division 5.1 (oxidizers)

- Division 6.1 (poisons)
- Hazard Class 8 (corrosives)
- Hazard Class 9 (miscellaneous)

Packages moving under this exception may not exceed 452 pounds (205 kilograms) gross weight.

Reuse of packaging

Used packaging may be reused once for the shipment of hazardous waste to a designated facility without being subject to the inspection and testing requirements for the reuse and reconditioning of packaging if:

- The waste is transported by highway.
- The waste is offered for transport within 24 hours after it is packaged for transport.
- The package is free from leaks.
- The package is loaded by the shipper and unloaded by the consignee, unless the motor carrier is a private or contract carrier.
- The remaining DOT Hazmat requirements are satisfied (DOT, § 173.12(c)).

Agricultural Operations

Formulated agricultural chemicals are exempt from marking requirements and certain packaging requirements when offered for transportation in less-than-case lot quantities, or when repackaged if:

- The inside packaging is enclosed in strong outside packagings and cushioned if necessary.
- Each inside packaging does not exceed 2.6-gallon (10-liter) capacity for liquids or 33 pounds (15 kilograms) for dry materials.
- The gross weight does not exceed 110 pounds (50 kilograms) in each vehicle.
- The shipment is moving by private motor carrier to an ultimate point of application.
- The distance does not exceed 100 miles. (DOT, § 173.5).

Other Exceptions

There are a number of other exceptions available under the regulations, which include the following:

- Chemical kits (kits containing limited quantities of corrosive liquids in inner receptacles of not over six fluid ounces each (DOT, § 173.161)
- Paint, paint-related material, adhesives, and ink (DOT, § 173.173)
- Lithium batteries and cells (DOT, §173.185)
- Matches (DOT, § 173.186)

7.9   TRAINING REQUIREMENTS

DOT's hazardous-materials transportation regulations require all persons involved in the transportation of hazardous materials to receive three types of training: general-awareness training, function-specific training, and safety training (DOT, §172.700). This training applies to anyone who transports hazardous materials; completes the shipping papers; handles, packages, marks, labels, loads, or unloads hazardous materials; or manufactures, reconditions, or tests hazardous-materials packaging.

General-awareness training is designed to make all employees aware of the various regulatory provisions and how to use them. DOT does not specify the method for conducting such training.

Function-specific training requires institutions to assess the particular job functions performed by each employee. Each employee then must receive training so that he or she can perform job functioning in compliance with the regulations. For instance, an employee whose job responsibilities include completing the shipping documentation must receive training in how to complete the shipping paper properly for the materials shipped by the company. An employee who loads and unloads material must be given instructions on proper handling, labeling, marking, and placarding requirements.

Safety training requires all hazardous-materials employees to be trained in the emergency-response measures for the material at issue. This training must include:

- emergency-response information
- measures to protect employees from work-place exposure hazards
- methods and procedures for avoiding accidents, such as the proper procedures for handling packages containing hazardous materials

Hazard-communication programs required by OSHA and EPA meet DOT requirements, so long as the employer's programs cover the listed areas.

All employees must be tested in each area to ensure comprehension and to determine if they can perform their job function in compliance with the regulations. The regulations do not specify the method of testing. However, in testing for general awareness, a written test is generally the preferred method of testing. Employees who fail to obtain a perfect score should be retrained and retested on the areas of difficulty until they understand their mistakes. The educational or healthcare institution should give employees a copy of their test so that they can use it for review. Employees also should be tested on their particular job responsibilities to ensure proper function-specific training and safety training.

The institution must maintain with the personnel files for that individual a record of the training that includes:

- employee's name
- most recent training date
- description, copy, or location of training materials
- name and address of person providing the training
- certification that the employee has been trained and tested

The certification may be a document from a training company indicating that the employee participated in the program or that training was conducted on the institution's premises. It can also be a statement by a supervisor that the employee was trained on that day and passed the relevant testing. It is not necessary to keep a copy of the employee test results as part of record keeping.

These records must cover the training over the preceding two-year period and must be retained for as long as that employee is employed and for 90 days thereafter. All employees must be trained within 90 days of employment or a change of job function and retrained at least every two years. During the interim time prior to satisfactorily completing training, all employees must perform their duties under the supervision of a properly trained and knowledgeable employee.

7.10 REGISTRATION

Shippers and carriers of certain types and quantities of hazardous materials are required to register with the U.S. DOT and pay a registration fee of $300 each year, an amount which may be increased in 1996 (DOT, §§107.601–.620). This registration fee applies to shipments of the following materials in foreign (moving through the United States), interstate, or intrastate commerce:

- any quantity of highway route–controlled radioactive materials
- more than 55 pounds (25 kilograms) of certain explosive materials
- more than 1.06 quarts (1 liter) per package of a material classified as extremely toxic by inhalation
- a hazardous material in a bulk packaging having a capacity equal to or

greater than 3,500 gallons for a liquid or gas (13,230 liters) or more than 468 cubic feet (13,254 cubic meters) for solids
- shipments in other than bulk packaging of 5,000 pounds (2,268 kilograms) gross weight or more of one class of hazardous materials for which placarding is required

The registrant will receive a Certificate of Registration which must be maintained at the principal place of business for a period of three years from the date of issuance of the Certificate along with the registration statement filed with DOT. Motor carriers also are required to carry on board their vehicle the Certificate of Registration or another document bearing the registration number as follows: "U.S. DOT Hazmat Reg. No. _____."

7.11   REPORTING TRANSPORTATION INCIDENTS

Carriers also have the responsibility of reporting incidents involving hazardous materials to the DOT or other responsible agency, depending on the nature of the incident. There are two separate reporting requirements under DOT's hazardous-materials regulations—immediate reporting of an incident resulting in death or serious injury to a person, or significant threat to the public health and/or environment; and detailed incident reporting within 30 days of the event (DOT, §§ 171.15, 171.16).

A carrier is required to report at the earliest practicable moment an incident involving any of the following:

- death or serious injury (hospitalization)
- damage to the carrier or property exceeds $50,000
- evacuation of the general public lasting one or more hours
- closure of one or more major transportation arteries or facilities for one or more hours
- fire, breakage, spillage, or suspected contamination involving radioactive materials in transport
- fire breakage, spillage, or suspected contamination occurs involving the shipment of etiologic agents
- release of a marine pollutant in a quantity exceeding 119 gallons (449 liters) for liquids or 882 pounds (400 kilograms) for solids
- a situation not satisfying any of the above, but is of such a nature that the carrier believes there is a continuing danger to life at the scene

This notice must be given either to DOT or, for a release or suspected release of etiologic agents, to the Director of the Center for Disease Control in Atlanta, Georgia. This notice must consist of the following information:

- name of person reporting incident
- name and address of carrier represented by individual reporting incident
- phone number of reporter
- date, time, and location of incident
- extent of injuries, if any
- classification, name, and quantity of hazardous material involved
- type of incident and nature of danger posed by hazardous material

A separate incident notice must be submitted to DOT by the carrier within 30 days of the incident for most materials (DOT, §171.16). The details of the incident must be reported on DOT Form F 5800.1. If the release involves a hazardous waste, a copy of the hazardous-waste manifest must be attached to the report and an estimate of the quantity of waste removed from the scene, the name and address of the facility to which the waste was taken, and the method of disposal of any removed waste must be entered on the form.

This detailed incident report must be made to the Information Systems Manager at the Research and Special Programs Administration of DOT. A copy of the report must be retained by the carrier for a period of two years at the carrier's principal place of business.

7.12   ENFORCEMENT

DOT may assess both civil and criminal penalties for violations of its hazardous materials regulations. Generally, DOT fines the responsible companies rather than individual employees. However, the governing statute allows DOT to impose penalties on individuals as well if it determines that an employee's conduct is particularly egregious. Civil penalties may not exceed $25,000 per violation. Criminal penalties may include a fine and up to five years' imprisonment.

REFERENCES

(DOT means U.S. Department of Transportation)

DOT. 1994. Proposed Rule. Federal Register (Dec. 21, 1994).
DOT. 1993. Proposed Rule. Federal Register (July 1993).

49 United States Code § 1802(15).
DOT. Proposed Rule. Federal Register 58:36920 (July 9, 1993).
DOT. Proposed Rule. Federal Register 59:65860 (Dec. 21, 1994).
DOT, 49 Code of Federal Regulations Parts 173, 174, 175, 176, 178.
49 Code of Federal Regulations § 171.8.
49 Code of Federal Regulations § 171.15.
49 Code of Federal Regulations § 171.16.
49 Code of Federal Regulations § 172.101(c)(10)(i).
49 Code of Federal Regulations § 172.200–172.205.
49 Code of Federal Regulations § 172.202.
49 Code of Federal Regulations § 172.203.
49 Code of Federal Regulations § 172.204(a)(1).
49 Code of Federal Regulations §§ 172.300–172.338.
49 Code of Federal Regulations § 172.301.
49 Code of Federal Regulations § 172.302.
49 Code of Federal Regulations § 172.310.
49 Code of Federal Regulations § 172.312.
49 Code of Federal Regulations § 172.313.
49 Code of Federal Regulations § 172.316.
49 Code of Federal Regulations § 172.320.
49 Code of Federal Regulations § 172.322.
49 Code of Federal Regulations § 172.324.
49 Code of Federal Regulations § 172.325.
49 Code of Federal Regulations §§ 172.326–172.331.
49 Code of Federal Regulations §§ 172.400–172.407.
49 Code of Federal Regulations § 172.402.
49 Code of Federal Regulations § 172.506.
49 Code of Federal Regulations § 172.508.
49 Code of Federal Regulations § 172.602.
49 Code of Federal Regulations § 172.700.
49 Code of Federal Regulations § 173.2a(b).
49 Code of Federal Regulations § 173.4.
49 Code of Federal Regulations § 173.5.
49 Code of Federal Regulations § 173.12.
49 Code of Federal Regulations § 173.12(b).
49 Code of Federal Regulations § 173.22.
49 Code of Federal Regulations § 173.28.
49 Code of Federal Regulations § 173.161.
49 Code of Federal Regulations § 173.173.
49 Code of Federal Regulations § 173.185.
49 Code of Federal Regulations § 173.186.
49 Code of Federal Regulations § 178.503.
49 Code of Federal Regulations § 383.

Chapter *8*

Waste Management and Disposal

8.1 INTRODUCTION AND BASIC FUNCTIONS

Medical and educational institutions generate a broad spectrum of wastes. Nonhazardous solid waste streams include paper and paperboard, yard wastes, glass, metals, plastics, and food wastes. These same facilities may generate a broad range of hazardous wastes, but in smaller quantities. A large university may dispose of over 7,000 containers holding more than 2,100 different chemicals and mixtures. (U.S. EPA 1990a). EPA estimates that research and educational institutions generate up to 4,408 tons of hazardous waste per year. You may derive some comfort from knowing that this figure is less than 1% of the national total of hazardous waste generated each year.

In this chapter, we explain the "a b c's" of how to traverse the regulatory landscape of solid waste. Congress enacted the Resource, Conservation and Recovery Act (RCRA) in 1976 to provide a framework for regulating the generation, transport, and disposal of solid and hazardous waste (O'Reilly, Kane, and Norman 1994). The Comprehensive Environmental Response, Compensation, and Liability Act (CERCLA), or "Superfund," was subsequently passed by Congress, to finance the cleanup of hazardous waste at inactive disposal facilities (USC § 9601 et. seq.). CERCLA also created stringent reporting obligations when hazardous materials are released. State laws and regulations should also be consulted because the rules at the state level may vary from federal requirements.

The facility manager dealing with the challenge of RCRA requirements for hazardous-waste movement is well advised to get the assistance of experienced consultants. Penalties are very large and very difficult to defend against (O'Reilly, Kane, and Norman 1994, Weiner and Bell 1994), so caution and sound advice are imperative.

Basic Functions

Simply put, regulatory managers must: (1) determine whether and where their institution generates solid waste, (2) characterize the waste as hazardous or non-hazardous, and (3) manage and dispose of the waste according to its characterization. Other than recycling requirements, ordinary solid waste is subject to few restrictions and can be sent to a municipal landfill or readily handled by a waste-disposal company. In contrast, hazardous waste is subject to a comprehensive regulatory scheme.

8.2   CHARACTERIZING AND MANAGING YOUR WASTE STREAMS

Waste-management requirements call for hospital and academic institutions to focus on appropriate hazardous-waste practices as a management goal and require a commitment of resources. Most large institutions have formal programs in place for addressing radioactive and hazardous-waste disposal. However, a 1987 EPA survey found several school districts studied had no budget for dealing with hazardous materials. Even institutions with programs in place must reevaluate their system from time to time to keep up with new developments, new waste-management practices, and waste-minimization strategies.

In educational institutions, chemistry departments tend to be the largest hazardous-waste generators. The largest waste stream associated with an academic institution is usually organic solvents. The second largest is often corrosives, which include organic and inorganic acids and bases. Metals (e.g., silver), unused chemicals, reaction products from experiments, and waste oil (from vacuum pumps and other rotating equipment) are other common wastes.

Secondary and vocational schools also generate waste in science laboratories, art and shop classes, vocational programs, and maintenance operations. Table 8–1 provides a list of high-school-course hazardous-waste generation (USEPA 1990a).

The first step for any institution seeking to comply with RCRA regulations is to conduct a chemical inventory. We have included sample inventory sheets (see Figure 8.1) to assist environmental professionals take inventory of the various departments for which they are responsible. After identifying potential waste streams, the next step is to determine how the wastes are being disposed. Current disposal practices should be compared with EPA and state requirements. Any practices in conflict with current requirements should be brought up to speed. Practices to be identified and avoided include such things as pouring wastes from chemistry experiments down the sink unless neutralized, or leaving and forget-

Table 8–1. High-School-Course Hazardous-Waste Generation

Courses Likely to Generate Hazardous Waste	Types of Wastes That Are Potentially Hazardous
Agricultural Arts	
Agriculture	Pesticides, fertilizers
Horticulture/Landscaping	Pesticides, fertilizers
Graphic Arts	
Art	Oil-based paints, solvents
Graphics	Inks, solvents, acids
Jewelry and Metalwork	Acids
Pottery and Ceramics	Metals in glazes, silica in clays
Painting/Drawing/Design	Oil-based paints, inks, solvents
Photo/Film Making	Silver, developing and fixing chemicals
Industrial Arts	
Carpentry/Woodworking	Stains, solvents, wood preservatives
Leather/Textiles/Upholstery	Dyes
Plastics	Ketones
Photography	Silver, photochemicals,
Printing/Photo/Graphics	inks, and solvents
Metalworking/Foundry	Metal dust
Welding	Metal waste
Auto Mechanics	Degreasing solvents,
Power/Auto Mechanics	oil, grease
Science Courses	
Natural Science	Various chemicals, acids, bases
Biology	
Chemistry	
Vocational Courses	
Trades and Industry	
Graphic Arts	Inks, solvents
Printing/Lithography	
Textile/Leather Products	Dyes
Body and Fender Mechanics	Batteries, paints, degreasing solvents, oil
Automobile Mechanics	Grease, acids, alkaline waste
Masonry	Paint, solvents, muriatic acid
Carpentry	Stains, solvents, paints, wood preservatives
Woodworking 1st year	
Woodworking—Advanced	
Machine Shop	Stripping and cleaning solutions, plating
Sheet Metal	bath residues, acids, bases, metal dust
Metalworking	
Welding and Cutting	Various chemicals
Cosmetology	
Health Courses	
Allied Health	
Laboratory/Chemical Technology	Various chemicals and pharmaceuticals
Nursing	

Source: US EPA

Firm _____	Waste Minimization Assessment	Prepared By _____
Site _____		Checked By _____
Date _____	Proj. No. _____	Sheet __ of __ Page __ of __

WORKSHEET 1	WASTE GENERATION: Questionnaire

Are facility-wide material balances routinely performed?   ❑ yes   ❑ no
Are they performed for each material of concern (e.g. solvent) separately?   ❑ yes   ❑ no
Are records kept of individual wastes with their sources of origin and eventual disposal?   ❑ yes   ❑ no
(This can aid in pinpointing large waste streams and focus reuse efforts.)

Having this type of data is important for the following reasons:
- the data define the scope of waste generation for the entire campus and for each department;

- realistic waste reduction goals can be established

- specific generators can be targeted for waste reduction; and

- costs for proper waste management can be determined.

If answer is No: It is recommended that methods for quantifying the waste generated for the entire campus and for each department be implemented. If adequate waste generation data are not available, establish an internal manifest system to be completed by each waste generator. An example of an internal manifest can be found on Worksheet 2. These forms should be kept on file and if possible, stored on a computer data base. Quarterly and yearly totals for hazardous waste generation can easily be determined using these manifests.

If answer is Yes: Establish campus-wide and departmental waste reduction goals. Setting specific goals provides an incentive to meet established goals. A committee should establish goals. Such committee should be made up of personnel from the campus environmental/safety office, administration, and professors/instructors from each waste generating department. Reduction goals should range from a 3% to 10% per year. The committee should meet quarterly to assess progress in achieving goals.

Figure 8.1. U.S. EPA waste-generation questionnaire.

Firm _____	Waste Minimization Assessment	Prepared By _____
Site _____		Checked By _____
Date _____	Proj. No. _____	Sheet __ of __ Page __ of __

WORKSHEET 2	WASTE GENERATION: Internal Manifest

Complete all information requested below. Do not leave any space or column empty (except Other Comments column)

Department_____

Person Completing Manifest _____ Phone No. _____

Date Generated	Chemical	Quantity	Gas, Liq., Solid	Hazard	Other Comments

Contact the campus Hazardous Waste Coordinator at_____
when manifest is completed and/or wastes need to be picked up.

Figure 8.1. (Continued)

Firm _____	Waste Minimization Assessment	Prepared By _____
Site _____		Checked By _____
Date _____	Proj. No. _____	Sheet __ of __ Page __ of __

| WORKSHEET 3 | WASTE MANAGEMENT: Laboratories |

This section is to be completed by the chemistry, biology, geology, physiology, physics, and any other departments that use chemicals in laboratories. Each question below is to be completed for each department:

Department person in charge of chemical storage/stock rooms: _____
Chairman of the department: _____
Total number of laboratories in department: _____
Number of research laboratories: _____
Number of undergraduate teaching laboratories: _____
Number of chemical storage (stock) rooms: _____
Number of professors in the department: _____
Subdivisions within the department (i.e., for chemistry department: general chemistry,
organic chemistry, analytical chemistry, etc.) _____
_____

Rank the subdivisions from the highest to the lowest for quantity of waste generated._____
_____
_____

How are chemicals purchased within the department?_____
_____
_____

Is there an oversight mechanism capable of monitoring all purchases of chemicals?_____
_____

Describe the current method of maintaining an active inventory of chemicals in stock: _____
_____
_____

Who in the department maintains Material Safety Data Sheet (MSDS) files?_____
_____
_____

How are wastes currently collected for disposal within the department?_____
_____
_____

Note: If you have trouble answering any of the above questions, investigate to find the answers. The answers will
assist in implementing waste reduction opportunities.

Figure 8.1. (Continued)

Firm _____	**Waste Minimization Assessment**	Prepared By _____
Site _____		Checked By _____
Date _____	Proj. No. _____	Sheet __ of __ Page __ of __

WORKSHEET **4A**	**WASTE MANAGEMENT:** **Science Departments**

The following checklist should be completed by each professor/instructor who supervises student laboratory exercises, supervises a research laboratory, or any staff person involved with handling chemicals including chemical stockroom supervisor, solutions preparation, technical supervision.

Department _____
Name of Person Completing this Checklist _____ Title _____

A. GOOD OPERATING PRACTICES

Are all affected personnel provided with detailed operating manuals or instruction sets?	❏ yes	❏ no
Are regularly scheduled training programs related to waste minimization?	❏ yes	❏ no
Are there employee/student incentive programs offered to all personnel?	❏ yes	❏ no
Does the facility have an established waste minimization program in place?	❏ yes	❏ no
If yes, is a specific person assigned to oversee the success of the program?	❏ yes	❏ no

Discuss the goals of the program and results: _____
_____
_____

Has a waste minimization assessment been performed at the facility in the past? If yes, discuss: _____
_____
_____

B. MATERIALS HANDLING

Has a centralized purchasing program been established?	❏ yes	❏ no
Does the current program adequately prevent the generation of waste due to over-purchasing?	❏ yes	❏ no

Since a significant portion of laboratory waste is actually surplus reagent chemicals,

is it possible to purchase smaller quantities of reagent chemicals?	❏ yes	❏ no
Is it possible to increase the amount of sharing of chemicals between research laboratories?	❏ yes	❏ no

This would reduce the amount of surplus chemicals that require disposal.

Is obsolete raw material returned to the supplier?	❏ yes	❏ no
Is inventory used in first-in, first-out order?	❏ yes	❏ no
Is the inventory system computerized?	❏ yes	❏ no
Does the current inventory control system adequately prevent waste generation?	❏ yes	❏ no

What information does the system track? _____
_____
_____
_____
_____

Figure 8.1. (Continued)

Firm _____	Waste Minimization Assessment	Prepared By _____
Site _____		Checked By _____
Date _____	Proj. No. _____	Sheet __ of __ Page __ of __

| WORKSHEET **4B** | WASTE MANAGEMENT: Science Departments |

Is there a formal personnel and student training program on raw material handling, spill prevention, proper storage techniques, and waste handling procedures?   ❑ yes   ❑ no

Does the program include information on the safe handling of the types of drums, containers and packages received?   ❑ yes   ❑ no

How often is training given and by whom? _____

_____

_____

C. LABORATORY PRACTICES

Is it possible to reduce the volumes of reactants used in certain laboratory experiments without affecting the desired results?   ❑ yes   ❑ no

Instrumental methods use significantly smaller quantities of chemicals than wet chemistry methods. Is it possible to increase the use of instrumental analyses for selected experiments?   ❑ yes   ❑ no

Is it possible to substitute less hazardous chemicals in certain laboratory experiments such as: using sodium hypochloride for sodium dichromate, alcohols instead of benzene, cyclohexane for carbon tetrachloride, stearic acid for acetoamide, and any other potential substitutes?   ❑ yes   ❑ no

Is it possible to substitute specialty detergents for chromic/sulfuric acid for cleaning glassware?   ❑ yes   ❑ no

For certain undergraduate laboratory exercises, is it possible to pre-weigh chemical reactants for students? This would eliminate chemical waste due to spillage during weighing and transfer operations, by students.   ❑ yes   ❑ no

If solvents are used for cleaning, is counter current cleaning possible? (Using spent solvent for initial cleaning and fresh solvent only for the final cleaning.) This decreases the amount of reagent solvent used.   ❑ yes   ❑ no

Is a solvent sink used? If not, could one be used?   ❑ yes   ❑ no

Can solvent waste be redistilled and reused for classroom experiments or as thinners or degreasers by the maintenance department?   ❑ yes   ❑ no

If onsite solvent distillation is done, does it comply with fire and worker safety regulations?   ❑ yes   ❑ no

Are many different solvents used for cleaning?   ❑ yes   ❑ no

If too many small-volume solvent waste streams are generated to justify on-site distillation, can the solvent used for cleaning be standardized?   ❑ yes   ❑ no

Are all chemicals containers properly labeled?   ❑ yes   ❑ no

Are all wastes properly segregated?   ❑ yes   ❑ no

Has off-site reuse of wastes through Waste Exchange services been considered?   ❑ yes   ❑ no

Or reuse through commercial brokerage firms?   ❑ yes   ❑ no

If yes, describe results: _____

_____

_____

Can waste chemicals be destroyed, neutralized or treated to reduce hazards as the final step of selected laboratory classwork and research experiments?   ❑ yes   ❑ no

Figure 8.1.   (Continued)

Inst. _____	Waste Minimization Assessment	Prepared By _____
Site _____	Proc. Unit/Oper. _____	Checked By _____
Date _____	Proj. No. _____	Sheet __ of __ Page __ of __

WORKSHEET 5A	OPTION GENERATION: Science Departments

Meeting format (e.g., brainstorming, nominal group technique) _____

Meeting Coordinator _____

Meeting Participants _____

Suggested Waste Minimization Options	Currently Done Y/N?	Rationale/Remarks on Option
A. Good Operating Practices		
Establish waste minimization policy		
Set goals for source reduction		
Set goals for recycling		
Conduct annual assessments		
Provide operating manuals/instructions		
Employee/student training		
Increased supervision		
Provide employee/student incentives		
B. Materials Handling		
Centralize purchasing		
Purchase smaller quantities		
Share surplus chemicals		
Return material to supplier		
Minimize inventory		
Computerize inventory		
Formal training		

Figure 8.1. (Continued)

Inst. _____	**Waste Minimization Assessment**	Prepared By _____
Site _____	Proc. Unit/Oper. _____	Checked By _____
Date _____	Proj. No. _____	Sheet __ of __ Page __ of __

WORKSHEET 5B

┌─────────────────────────────────┐
│ **OPTION GENERATION:**   │
│ **Science Departments**   │
└─────────────────────────────────┘

Meeting format (e.g., brainstorming, nominal group technique) _____
Meeting Coordinator _____
Meeting Participants _____

Suggested Waste Minimization Options	Currently Done Y/N?	Rationale/Remarks on Option
C. Laboratory Practices		
Scale down experiments		
Increase instrument use		
Eliminate toxic chemical use		
Pre-weigh chemicals		
Standardize solvents & recycle		
Properly label containers		
Segregate wastes		
Recycle through waste exchange		
As final step, treat waste		

Figure 8.1.   (Continued)

Firm _____	**Waste Minimization Assessment**	Prepared By _____
Site _____		Checked By _____
Date _____	Proj. No. _____	Sheet __ of __ Page __ of __

WORKSHEET **6A**	**WASTE MANAGEMENT:** **Other Departments**

A. ART, THEATER ARTS, SCENERY SHOP, AND PRINTING

Is it possible to significantly reduce or eliminate use of oil-based paints?	❏ yes	❏ no
When spray painting, is it possible to use the following techniques to reduce the amount of paint used?	❏ yes	❏ no
- employ high transfer efficiency guns?	❏ yes	❏ no
- overlap the spraying pattern by 50%?	❏ yes	❏ no
- maintain a distance of 6 to 8 inches from the work piece?	❏ yes	❏ no
- hold the gun perpendicular to the surface?	❏ yes	❏ no
- trigger the gun at the beginning and end of each stroke?	❏ yes	❏ no
Use fully enclosed gun cleaning stations?	❏ yes	❏ no
Reuse clean-up solvent as thinner in next compatible batch of paint?	❏ yes	❏ no
Hazardous chemicals are used in art department subdivisions including silk screening, metal work, and sculpture. Is it possible to reduce or make substitutions for specific hazardous chemicals in any of these areas?	❏ yes	❏ no

If yes, discuss: _____

_____

_____

Certain photoprocessing cleaning chemicals are hazardous (e.g. chromic acid). Is it possible to substitute less hazardous cleaning compounds?	❏ yes	❏ no
In photographic processing is there currently a silver recovery unit in place to recover silver salts in the waste water?	❏ yes	❏ no
If there is no silver recovery unit, is it possible to install one?	❏ yes	❏ no
Is off-site recovery feasible?	❏ yes	❏ no
In pottery making or other related work, is it possible to eliminate use of lead glazes?	❏ yes	❏ no

B. MAINTENANCE SHOP

Is it possible to eliminate use of oil-based paints and replace with water-based paints?	❏ yes	❏ no
Is it possible to standardize oils used for many kinds of machinery?	❏ yes	❏ no
Can water-based cutting fluids be used in place of oil-based fluids?	❏ yes	❏ no
Does the facility have a proper coolant management program in place?	❏ yes	❏ no
Can hazardous solvent degreasers be replaced by alkaline degreasers or less hazardous solvent degreasers?	❏ yes	❏ no
If a vapor degreasing unit is used, is it always covered when not in use to reduce loss of solvent to the atmosphere?	❏ yes	❏ no
When using a vapor degreasing unit, are the parts rotated before removal to allow condensed solvent to return to the degreasing unit?	❏ yes	❏ no
Is it possible to restrict the number of parts that must be degreased to only those parts that badly need degreasing rather than routinely degreasing all parts?	❏ yes	❏ no

Figure 8.1. (Continued)

Firm _____	Waste Minimization Assessment	Prepared By _____
Site _____		Checked By _____
Date _____	Proj. No. _____	Sheet __ of __ Page __ of __

| WORKSHEET **6B** | **WASTE MANAGEMENT:** Other Departments | |

B. MAINTENANCE SHOP (CONT.)

To conserve use of reagent solvents, can dirty solvent be used for initial cleaning, and fresh
solvent used for final cleaning? ☐ yes ☐ no

Is a solvent sink being used? If not, could one be used? ☐ yes ☐ no

Is a bench-top still appropriate? If one is being used, does it comply with fire and worker
safety regulations? ☐ yes ☐ no

When spray painting, is it possible to use the following techniques to reduce
the amount of paint used?
- overlap the spraying pattern by 50% ☐ yes ☐ no
- maintain a distance of 6 to 8 inches from the work piece
- hold the gun perpendicular to the surface
- trigger the gun at the beginning and end of each stroke? ☐ yes ☐ no

For pesticide spraying equipment, can the generation of pesticide contaminated rinse water
be reduced or eliminated? This can be done by saving and using the rinse water for makeup ☐ yes ☐ no
of the next pesticide application solution.

To reduce or eliminate the need for spraying pesticides can either of the following be
implemented:
- irrigation injection where the pesticide formulation is injected directly into the
  sprinkler/irrigation system at a controlled rate and with adequate backflow
  prevention devices ☐ yes ☐ no
- spread pesticides in the dry powder form and then water them into the ground? ☐ yes ☐ no

For spent fluorescent lamps and mercury recovered from lab sink traps has an outside vendor
been contacted to investigate the feasibility of recovering mercury? ☐ yes ☐ no

Discuss any other methods used to minimize waste: _____

_____

_____

_____

_____

Figure 8.1. (Continued)

Inst. _____	Waste Minimization Assessment	Prepared By _____
Site _____	Proc. Unit/Oper. _____	Checked By _____
Date _____	Proj. No. _____	Sheet __ of __ Page __ of __

WORKSHEET **7**	**OPTION GENERATION:** **Other Departments**

Meeting format (e.g., brainstorming, nominal group technique)_____
Meeting Coordinator _____
Meeting Participants _____

Suggested Waste Minimization Options	Currently Done Y/N?	Rationale/Remarks on Option
A. Art, Theater Arts, Scenery Shop, and Printing		
Eliminate oil-based paint use		
Proper spray paint techniques		
Enclosed spray gun cleaning		
Reuse clean-up solvent		
Use less hazardous cleaners		
Recover photographic silver		
Eliminate use of lead-based glaze		
B. Maintenance Shop		
Eliminate oil-based paint use		
Proper spray paint techniques		
Standardize machine oil		
Use water-based cutting fluids		
Proper coolant management program		
Replace solvent degreasers		
Keep degreaser covered		
Operate degreaser properly		
Recycle pesticide rinse water		
Use dry pesticide or irrigation injection		
Recover mercury		

Figure 8.1. (Continued)

ting chemicals and used and unused reagents in unmarked test tubes, beakers, vials, or bottles.

A cradle-to-grave inventory-tracking program is critical for the environmental professional to keep up with current regulatory requirements for managing hazardous waste. The program should include plans for leftover chemicals. Centralizing waste management in one personnel position allows for efficient implementation of regulations and institution policy. Appointing a safety or waste-management officer for each department or for the entire school is a simple way to centralize and organize waste-management responsibilities and helps to ensure that institutional policies and waste-reduction goals are met. A training program for professional staff and students should be developed in conjunction with an institutional waste-management program.

8.3   SOLID WASTE IS BROADLY DEFINED
TO INCLUDE NONSOLIDS

What is meant by the term "solid waste"? RCRA broadly defines "solid waste" as:

> Any garbage, refuse, or sludge from a waste-treatment plant, water-supply treatment plant, or air-pollution-control facility, and other discarded material, including solid, liquid, semisolid, or contained gaseous materials resulting from industrial, commercial, mining, and agriculture activities and from community activities, but does not include solid or dissolved material in domestic sewage, or solid or dissolved materials in irrigation return flows or industrial discharges which are point sources subject to permits under section 402 of the Federal Water Pollution Control Act, as amended, or source, special nuclear, or by-product material as defined by the Atomic Energy Act of 1954, as amended (68 Stat. 923) (USC, § 6903(27)).

This definition is so broad it even encompasses liquid sludge and contained gaseous materials. EPA considers these to be "solid" wastes. This illustrates the crucial differences—and the resulting complexities—between "regulatory" management of your garbage and "regular" concepts of what the average person would consider to be controlled under the definition of solid waste. It's not as easy as you might think to get rid of your trash.

At its inception, RCRA required EPA to approve the establishment of state solid-waste-management plans (USC, § 6941-6943(a)). EPA also regulates the minimum requirements for landfills and the closing or upgrading of open dumps, incinerators, and waste-to-energy facilities. Solid-waste-disposal facilities at the local level must meet federal operating requirements under Subtitle D of RCRA. For example, municipal incinerator regulations include control of air emissions, solid waste-stream planning, and ash-disposal requirements.

While facility-operating requirements are detailed at the federal level, the practical aspects of solid-waste disposal are addressed at the state level. What can and what cannot be disposed of at landfills and incinerators? Who can use these facilities and who cannot? Where can your solid waste be disposed of and how much? State requirements require close attention. For example, many states do not allow yard (and grounds) waste and other specific solid-waste streams—such as tires, batteries, and oil—to be disposed of in local landfills. Table 8–2 provides a summary of such disposal bans in key states.

8.4 HAZARDOUS WASTE REQUIRES MANAGEMENT

The hallmark of the RCRA hazardous-waste program is its cradle-to-grave approach to regulating hazardous wastes from generation to ultimate disposal (O'Reilly, Kane, and Norman 1994). Operations generating any solid waste must determine whether the waste is hazardous. If it is, the generator, as well as subsequent transporters and treatment/storage/disposal facilities, must carefully follow prescribed practices.

Universities, hospitals, and secondary schools are most likely to encounter regulation as a hazardous-waste generator. Many hospitals and academic institutions may be either "small-quantity generators" or "conditionally exempt small-quantity generators" under RCRA. As discussed in detail below, environmental professionals whose facilities fall in these categories must perform a hazardous-waste determination and comply with manifest, label, marking, record-keeping, and container management practices; emergency preparedness and prevention; and authorized disposal requirements. Federal RCRA permitting will not be required to manage and store hazardous waste up to 1,000 kilograms (2,200 pounds) (EPA.CFR. § 261.5(a)) or 6,000 kilograms (13,224 pounds) in a 180-day period (EPA.CFR. § 260.10) respectively. However, some states have no small-quantity exemption.

8.5 IDENTIFICATION AND CLASSIFICATION SCHEME

Determining whether or not a waste is "hazardous" under RCRA is the first step in proper hazardous waste management. There are two basic categories of hazardous waste: (1) solid wastes that are listed as hazardous by EPA or the relevant

Table 8–2. Disposal Bans

State	Lead-acid Batteries	Yard Waste	Unprocessed Tires	Used Oil	Large Appliances	Other
California	X					
Connecticut	X	XA		X		B
D.C.		X				
Florida	X	X	X	X	X	C
Georgia	X					
Hawaii	X					
Illinois	X	X				
Iowa	X	X	X	X	X	D
Kansas			X			
Kentucky	X					
Louisiana	X		X	X	X	
Maine	X					
Michigan	X					
Minnesota	X	X	X			E
New Hampshire	X					
New Jersey	X					
New York	X					
North Carolina	X	X	X			
Ohio	X					
Oklahoma						
Oregon	X					F
Pennsylvania	X					
Rhode Island						
Virginia	X					
Washington	X					
Wisconsin	X	X	X		X	G
Wyoming	X					

A. Yard-waste disposal bans only apply to leaves.
B. Nickel-cadmium batteries.
C. Construction & demolition debris.
D. Nondegradable grocery bags; beverage containers returned to wholesalers through the state's mandatory deposit law.
E. Dry-cell batteries that contain mercuric oxide or silver-oxide electrodes, nickel-cadmium, or sealed lead-acid. Mixed unprocessed waste in metro area.
F. Recyclable material that has already been separated.
G. Aluminum, plastic, steel and glass containers, corrugated paper and paperboard, foam polystyrene packaging, magazines, newspaper, and office paper are banned from disposal unless municipalities are certified as having an "effective" source-separation program.

state agency and (2) solid wastes which, while not listed, display one of four hazardous "characteristics."

Listed Hazardous Wastes

EPA "listed" hazardous waste include:

- hazardous wastes from nonspecific sources (e.g., spent solvents)
- hazardous wastes from specific sources

Characteristic Hazardous Waste

Reference to the lists of hazardous wastes does not discharge a generator's obligation to identify and classify hazardous wastes. Rather, the environmental professional must evaluate waste streams to determine whether they might be hazardous by virtue of exhibiting a hazardous "characteristic." That determination may be based on knowledge of the hazardous characteristics of the waste or testing. The four characteristics that make a waste hazardous are:

- ignitability
- reactivity
- corrosivity
- toxicity

Each of these characteristics, as well as the proper methods for testing for their presence, is described below. But remember, new determinations are needed when processes or materials change.

The following rules apply to "characteristic" hazardous waste generally:

- An actual test for a characteristic is not necessary if an employee can determine from his or her knowledge of the waste that it cannot reasonably be expected to exhibit the characteristic.
- Whenever a test for a characteristic is not performed because of an employee's knowledge of the waste, a brief memorandum fully explaining the basis for the decision not to perform the particular test should be prepared and retained in the files.
- Tests for determining whether a waste exhibits a characteristic require use of a representative sample of the waste (40 EPA § 261, App. I).

The characteristic of *ignitability* was established to identify solid wastes capable of causing or exacerbating a fire during routine handling. The test for ignitability is most frequently expressed as whether the waste has a flash point of

140°F (60°C) or less. EPA, however, has expressed the test in somewhat more definite terms (EPA.CFR. § 261.21). In addition to the 140°F flash point, the presence of other properties causes a waste to exhibit the characteristic of ignitability.

The characteristic of *corrosivity* was established on EPA's premise that wastes that could corrode metal might escape from their containers and liberate other wastes. Corrosive wastes are (1) aqueous wastes with pH less than or equal to 2 or greater than or equal to 12.5, or (2) liquid wastes that corrode steel at a rate greater than 6.35 mm (0.25 inch) per year. EPA has designed test methods for identifying the characteristic of corrosivity (EPA.CFR. § 261.22).

EPA established the characteristic of *reactivity* in order to regulate wastes that were extremely unstable with the tendency to react violently or explode during handling. Several properties cause a waste to exhibit the characteristic of reactivity including:

- instability and readiness to undergo violent change without detonating
- violent reactions with water, potentially leading to explosiveness or vapors or fumes presenting a danger to human health or the environment
- capability of detonation or explosive reaction (EPA.CFR § 261.23)

A waste exhibits the characteristic of *toxicity* if an extract from a representative sample of the waste, using a test method specified by EPA, contains designated contaminants in concentrations equal to or greater than values established by EPA (EPA CFR § 261.24). The toxicity characteristic was developed in order to identify wastes that are likely to leach hazardous concentrations of specific toxic constituents into groundwater by mismanagement. Twenty-five organic constituents, four pesticides, two herbicides, and eight heavy metals are designated under EPA's toxicity characteristic rule. The test method specified by EPA for identifying the toxicity characteristic is the Toxicity Characteristic Leaching Procedure (TCLP) (EPA CFR § 261, Appendix II).

8.6 EXEMPTIONS

EPA's regulations on waste also automatically exempt certain solid wastes from being considered hazardous. Among those subject to exemption:

1. agricultural wastes that are returned to the ground as fertilizer
2. utility wastes from coal combustion
3. certain chromium-bearing wastes

Further, EPA has made provisions for some limited regulatory exemptions under narrowly defined circumstances. An example of such a circumstance may be a hazardous waste that is generated in a product or transport vehicle, pipeline, or manufacturing process unit prior to removal for disposal. EPA has also adopted a conditional exemption for waste samples to be used for testing.

8.7   MIXTURE RULES

EPA's rules for determining whether or not a mixture of a nonhazardous solid waste and a hazardous waste is itself a hazardous waste differ depending on whether the hazardous component of the mixture is a *listed* or *characteristic* hazardous waste (EPA § 261.3(a)(2)). The "mixture rule" does not apply where two wastes or waste streams are generated separately and the hazardous waste is mixed with the nonhazardous waste. The mixture of wastes in this manner would constitute "treatment," which requires a TSD operating permit and cannot be conducted by generators.

A mixture of a *listed* hazardous waste and a nonhazardous solid waste is a hazardous waste unless:

- The listed hazardous waste in the mixture was listed solely because it exhibited a hazardous characteristic (i.e., ignitability, corrosivity, reactivity, and toxicity) and the mixture does not exhibit that characteristic. Or,
- The mixture consists of specified listed hazardous wastes and wastewater, the discharge of which is subject to regulation under the Clean Water Act, if the concentration of the specified listed hazardous waste does not exceed certain concentrations.

In contrast to these somewhat complicated rules for mixtures of nonhazardous solid waste and *listed* hazardous wastes, the rules for mixtures of nonhazardous solid wastes and *characteristic* hazardous wastes is far simpler. A mixture of a characteristic hazardous waste and a nonhazardous solid waste is considered hazardous only if the entire mixture continues to exhibit the hazardous characteristic. This rule applies in situations where a hazardous chemical (which would be a hazardous waste if discarded) and nonhazardous substances are used in a process or as ingredients and the resulting waste stream contains the constituents.

Of equal importance is EPA's "derived from" rule. The derived from rule designates a solid waste as hazardous if it is generated from the treatment, storage, or disposal of a listed hazardous waste, unless exempted. The waste may become nonhazardous if the listed hazardous waste from which it is derived becomes delisted.

8.8   LARGE- AND SMALL-GENERATOR CLASSIFICATION

EPA regulations define generators as "any person, by site whose act or process produces hazardous waste [identified or listed] or whose act first causes hazardous waste to become subject to regulation." Generators should consult with their state agency to determine whether their states adhere to the federal classification scheme or go further.

Large-Quantity Generator

Facilities that generate more than 1,000 kilograms (2,200 pounds) of hazardous waste per month are considered "large quantity generators" under the federal RCRA scheme and must comply with the broadest range of legal requirements. The determination of large-quantity generator status is made on a facility-by-facility basis. EPA interprets the 1,000 kilograms (2,200 pounds) threshold to apply to the total of all nonacutely hazardous wastes generated at a facility. Therefore, if a university generates 300 kilograms (660 pounds) of hazardous waste X, 500 kilograms (1,100 pounds) of hazardous waste Y, and 300 kilograms (660 pounds) of hazardous waste Z in a given month, it is a "large-quantity generator."

Small-Quantity Generator

Under the federal RCRA program, generators of small quantities of hazardous waste, between 100 and 1,000 kilograms (2,200 pounds) of hazardous waste monthly, are subject to somewhat less rigorous requirements than are large-quantity generators. To qualify as a "small-quantity generator," the production of all hazardous wastes on a monthly basis must not exceed 1,000 kilograms. If the 1,000-kilograms threshold is exceeded, then all requirements for large-quantity generators are applicable. An institution may move in and out of the small-quantity generator category on a month-to-month basis. Any shift from large- to small-quantity-generator status must be documented.

The following are ways in which the federal RCRA rules are relaxed for small-quantity generators:

- Subject to certain restrictions, a small-quantity generator may accumulate hazardous waste twice as long as large-quantity generators (generally 180 days) before the waste must be sent to a licensed TSD facility.
- Under certain circumstances *where waste is sent off-site* for *reclamation,* small-quantity generators do not have to comply with RCRA manifest

requirements (although preparing a manifest is good practice to avoid questions in the future regarding the fate of waste).

- Small-quantity generators have a greater period of time in which to notify regulatory agencies that they have not received confirmation that their waste was delivered to the designated facility.

- Small-quantity generators do not have to submit biennial reports under the federal RCRA scheme.

- The small-quantity generator's certification of having made efforts toward waste minimization is different from the large-quantity generator's.

- A small-quantity generator does not have to comply with the rule that requires that containerized ignitable or reactive waste be stored 50 feet or more inside the facility's property line.

Conditionally Exempt Small-Quantity Generators

In addition to small-quantity generators, the federal RCRA program creates a class of generators that produce no more than 100 kilograms (220 pounds) of hazardous waste in a calendar month. This subclass of small-quantity generators is "conditionally exempt" from the RCRA program. "Conditionally exempt small-quantity generators" must determine whether any solid waste they produce is hazardous and ensure that the waste is sent to a licensed facility. If the conditionally exempt generator accumulates more than 100 kilograms of hazardous waste at one time, then it must comply with the rules applicable to small-quantity generators. If a conditionally exempt small-quantity generator generates more than one kilogram of *acutely* hazardous waste or a total of 100 kilograms of any residue or contaminated soil, waste, or other debris resulting from the cleanup of a spill of acute hazardous waste, then all quantities of that acutely hazardous waste are subject to regulation in accordance with the rules for large-quantity generators.

8.9   GENERATOR NUMBERS

Every hazardous-waste generator must register to obtain a hazardous-waste-generator identification (I.D.) number in order to ship hazardous waste to an authorized TSD facility (EPA § 262.12). Generators are required to use the I.D. number on all waste manifests. The use of I.D. numbers precludes anonymity of parties subject to the RCRA program (e.g., "midnight dumpers"). A generator that does not have an I.D. number risks fines and other penalties. In most cases, generators must only register with the proper state agency; however, some states require generators to register with EPA. The federal registration form is EPA

Form 8700-12. Many state agencies have based their registration form on EPA Form 8700-12.

8.10   ON-SITE ACCUMULATION AND MANAGEMENT

Under federal hazardous waste regulations, generators may accumulate and store hazardous waste on-site for a limited period of time. The fundamental rule is that large-quantity generators can store hazardous waste no more than 90 days in a central storage area (in either containers, generally drums, or tanks) before sending the waste off-site to a waste-management facility (EPA § 262.34(a)). (Storage at the point of generation—"satellite accumulation"—is discussed later in this section.) Storage of hazardous waste beyond the 90-day limit requires a permit to operate as a TSD facility (EPA § 262.34(b)).

The 90-day period begins to run the day the first waste is placed in the container or tank in the central storage area. All of the hazardous waste that the container or tank holds must be shipped for off-site management within the 90-day period, including the waste that is added on the 89th day.

The following standards apply to hazardous wastes stored in a central area during the 90-day accumulation period:

- Generators of hazardous wastes that are accumulated in drums and tanks must follow certain container-management rules.
- The date on which accumulation of wastes begins must be clearly marked and visible for inspection.
- Drums and tanks must be marked clearly with the words "Hazardous Waste."
- A personnel training program, preparedness and prevention plan, and a contingency and emergency response plan must be in place.

A small-quantity generator (one that generates greater than 100 but less than 1,000 kilograms (2,200 pounds) in a calendar month, may accumulate hazardous waste on-site for 180 days provided the generator:

- complies with the rules for container and tank management
- never accumulates hazardous waste in excess of 6,000 kilograms (13,220 pounds)
- properly records the date when accumulation begins, labels all hazardous-waste containers, and adheres to EPA's preparedness and prevention rules
- has in place the procedures for emergency responses (EPA § 262.34(d))

If the small-quantity generator has to send its waste to a TSD facility 200 or more miles away, then the small-quantity generator can accumulate hazardous waste for 270 days if it adheres to these same conditions (EPA § 262.34(e)).

There is no time limit on how long conditionally exempt small-quantity generators (those that generate no more than 100 kilograms (220 pounds) of hazardous waste in a calendar month) can accumulate hazardous waste. However, if the conditionally exempt generator accumulates, at one time, more than 100 kilograms (220 pounds) of hazardous waste, then the generator must comply with the rules for small-quantity generators described in the immediately preceding section (EPA § 262.34(f)).

8.11   SATELLITE ACCUMULATION

The "satellite accumulation" rule allows collection of hazardous waste at the point of generation before removal of the waste to a central storage area. If certain conditions are satisfied, a generator (regardless of its classification) can accumulate up to 55 gallons (207 liters) of hazardous waste, or one quart of acutely hazardous waste, at or near the point of generation—the so-called "satellite accumulation area" (EPA § 262.34(c)). As soon as 55 gallons of hazardous waste or one quart of acutely hazardous waste are accumulated in the satellite area, a generator has three days to move the waste to a central storage area. If more than 55 gallons of hazardous waste, or one quart of acutely hazardous waste, accumulate at the point of generation, the satellite area containers must be marked with the date the excess is accumulated.

Hazardous waste may be accumulated in satellite accumulation areas at the point of generation if:

- the waste is put in containers that are in good condition
- the containers are compatible with the waste
- the container is kept closed except when adding or removing waste
- the containers are marked "Hazardous Waste" or other words that identify their contents

Some states have not adopted the satellite accumulation rule. Generators should consult with their state agency to determine whether their state has adopted the rule. In states where the rule has not been adopted, the 90-day accumulation period begins to run as soon as the waste is generated and collected at the point of generation.

8.12   CONTAINER MANAGEMENT

RCRA imposes the following requirements on generators of hazardous waste that accumulate wastes in containers (drums) prior to shipment off-site for management:

- Containers must be in good condition and, if a container begins to leak, its contents must be transferred to a container in good condition.
- Containers must be made of or lined with materials that will not react with or are not otherwise incompatible with the waste to be stored in the container.
- Containers must be kept closed at all times except when adding or removing wastes. Before their shipment for off-site management, hazardous waste containers should be properly sealed, leaving three to five inches of top space for expansion.
- Containers must be handled so as to avoid ruptures or leaks.
- Containers must be inspected at least weekly for leaks or deterioration.
- Containers holding ignitable or reactive waste must be located at least 50 feet from the property line (small-quantity generators do not have to adhere to this rule) as well as shielded from sources of ignition or reaction such as open flames, smoking, radiant heat, etc.
- "No Smoking" signs must be posted in areas where ignitable or reactive wastes are stored.
- Incompatible wastes must not be placed in the same containers, and waste must not be placed in an unwashed container that previously held an incompatible waste. If the central storage area contains drums of incompatible waste, the drums must be separated by means of a dike, beams, or wall (EPA § 262.34(a)(1)(i), EPA § 265, Subpart I).

8.13   STORAGE-TANK MANAGEMENT

EPA rules for temporary on-site storage of hazardous waste in tanks or tank systems are stringent (EPA § 262.34(a)(1)(ii), EPA § 265, Subpart J—except § 265.197(c) and § 265.200). Those rules require the integrity of existing tanks to be assessed and will also require secondary containment systems capable of preventing releases to the ground or groundwater as well as release-detection mechanisms by 1995. State hazardous-waste regulations may add further requirements. We discuss storage-tank management in more detail in a later chapter.

8.14   HAZARDOUS-WASTE MANIFEST

A manifest must be prepared prior to the off-site shipment of hazardous waste. Manifests must accompany all off-site shipments of hazardous waste, and the responsibility for preparing the manifest falls upon the generator. This section provides guidance for general manifest requirements, manifest contents, routing manifests, and notice of receipt of the waste shipment at the treatment, storage, or disposal (TSD) facility.

Generators that transport, or offer for transportation, hazardous waste for off-site treatment, storage, or disposal must prepare and use a manifest form (EPA, § 262.20(a)). There are two exceptions to the manifest requirement. The first covers conditionally exempt small-quantity generators that do not have to complete manifests (EPA, § 261.5). The second covers some instances in which small-quantity generators send hazardous waste off-site for reclamation (EPA, § 262.20(e)).

Under the federal RCRA program, EPA Form 8700-22 and EPA Continuation Form 8700-22A (if necessary) are the required manifest forms. Copies of these forms and their instructions are obtainable from state waste officials. Some states require use of EPA's uniform manifest form; other states require use of their own manifest form. Generators should consult with their state agency to determine specific state-manifest requirements.

There are three parts to EPA and state-manifest forms, the first part of which must be completed by the waste generator. In the first part, the environmental professional who prepares and signs the manifest form designates the permitted TSD facility contracted to handle the waste identified by the manifest. Designation of the TSD facility is perhaps the most important aspect of the manifest system and should include the TSD facility's permit number. The environmental professional should also designate an alternate facility to receive the waste in the case of emergencies. The first part of the manifest also requires the following information:

- the manifest document number
- the name, address, telephone number, and EPA generator I.D. number
- the DOT name and number for the hazardous waste being shipped
- the total quantity of each hazardous waste being shipped by weight or volume
- the type and number of containers used in transporting the hazardous waste

The employee's signature on the manifest certifies the accuracy of the information on the manifest, the hazardous-waste minimization efforts, and compliance with RCRA's pre-transport requirements. In the second part of the manifest, the initial transporter (and any subsequent transporters) acknowledges receipt of the shipment, indicates the date received, and, in some states, provides its hazardous-waste transporter I.D. number. The third part of the manifest provides space for the TSD facility operator to sign and certify receipt of the shipment and the date received.

Section 3002(b) of RCRA requires large-quantity generators of hazardous waste to certify on their hazardous-waste manifests that they have a program in place to reduce the volume or quantity and toxicity of hazardous waste generated to the extent "economically practicable." The term "economically practicable" is to be defined and determined by the generator. Small-quantity generators, on the other hand, need only certify on manifests that they have made a good-faith effort to minimize their hazardous-waste generation. For both large- and small-quantity generators, the EPA manifest requires the following certification:

> I hereby declare that the contents of this consignment are fully and accurately described above by proper shipping name and are classified, packed, marked, and labeled, and are in all respects in proper condition for transport by highway according to applicable international and national government regulations.

Large-quantity generators must additionally certify that:

> I have a program in place to reduce the volume and toxicity of waste generated to the degree I have determined to be economically practicable and that I have selected the practicable method of treatment, storage, or disposal currently available to me which minimizes the present and future threat to human health and the environment.

Small-quantity generators, on the other hand, have a lesser certification obligation, and must certify that:

> I have made a good-faith effort to minimize my waste generation and select the best waste-management method that is available to me and that I can afford.

EPA has published interim guidelines to provide guidance to generators who must certify that they have a waste-minimization program in place (Federal Register 1993). The corporate officer responsible for ensuring RCRA compliance must sign the waste-minimization program.

The environmental professional should make at least the following copies of the manifest:

- one copy for his or her own records
- one copy for each transporter and TSD facility handling the hazardous waste, for their records
- one copy for the TSD facility operator to return to the facility as notification that the waste was received (EPA, § 262.23)

Some states also require that a copy of the manifest be sent to the appropriate state agency. For interstate shipments, both the generator's and TSD facility's state agency may request a copy. For further requirements on routing the manifest, generators should consult with their state agency.

Thirty days after the waste is shipped, the environmental professional should check to see if a copy of the manifest has been received from the TSD facility. If the receipt copy has not arrived within 30 days, the environmental professional should contact the transporter and the designated TSD facility. All contacts with the transporter and the designated TSD facility regarding shipments of hazardous waste should be documented.

An Exception Report must be filed if a large-quantity generator does not receive a copy of the manifest confirming the arrival of the waste shipment at the TSD facility within 45 days after the initial transporter accepted the waste (EPA CFR 262.42(a)). Small-quantity generators have 60 days after the initial transporter accepted the waste (EPA CFR § 262.20(b)). Exception Reports must include: (1) a legible copy of the manifest for which there is no confirmation of delivery to the TSD facility, and (2) a cover letter signed by the responsible employee, explaining the efforts taken to locate the hazardous-waste shipment and the results of those efforts (EPA CFR § 262.42(a)(2)). The Exception Report should be addressed to the appropriate EPA Regional Office or state agency office.

8.15 Land-Disposal Restriction Requirements

Both large- and small-quantity generators have another notice or notice-and-certification obligation at the time they manifest their hazardous waste. (Conditionally exempt small-quantity generators do *not* have this obligation.) This notice or notice-and-certification obligation arises from the "land-disposal restriction" requirements, which require treatment of hazardous waste using the best demonstrated treatment method before such waste is disposed of on land (EPA, § 268).

There are over one hundred pages of federal regulations setting out the land-disposal restriction requirements. In practice, the hazardous-waste service contractor with whom generators do business will take the steps necessary to ensure that facilities comply with the requirements. Nevertheless, the law imposes the land-disposal restriction notice or notice-and-certification requirement on the generator of hazardous waste. For that reason, and despite the fact that service contractors may, in practice, discharge the institution's duties, generators should be familiar with the requirements.

Land-disposal restriction requirements apply to both *listed* and *characteristic* hazardous waste. Generators subject to the land-disposal restriction requirements (i.e., all that are not conditionally exempt) must determine whether the wastes they are manifesting meet certain *treatment standards*. Treatment standards are expressed in terms of either (1) concentrations of a particular hazardous constituent in the hazardous waste, or (2) a specific treatment technology, for example, incineration.

If a generator determines that a hazardous waste does *not* meet an applicable treatment standard as set out in 40 C.F.R. Part 268, then it must include a notice to that effect with its manifest. The notice must contain the following information:

- the waste's EPA hazardous waste number
- the corresponding land-disposal restriction treatment standard, including the five-letter treatment code where treatment standards are expressed as specified technologies (see 40 EPA Part 268)
- the manifest number associated with the shipment of waste
- test analysis data, where available

If, on the other hand, the generator finds that the hazardous waste meets the applicable treatment standard as set out in 40 C.F.R. Part 268, then it must include a notice-and-certification to that effect with its manifest. The notice-and-certification must include the same information listed in the four items above, plus the following certification:

> I certify under penalty of law that I personally have examined and am familiar with the waste through analysis and testing or through knowledge of the waste to support this certification that the waste complies with the treatment standards specified in 40 C.F.R. Part 268 Subpart D and all applicable prohibitions set forth in 40 C.F.R. 268.32 or RCRA section 3004(d). I believe that the information I submitted is true, accurate, and complete. I am aware that there are significant penalties for submitting a false certification, including the possibility of a fine and imprisonment.

Whether a generator determines that its hazardous waste *does* or *does not* meet the relevant land-disposal restriction treatment standard, if the environmental professional has made this determination solely on the basis of its knowledge of the waste (rather than based upon analytical testing results), all supporting data

relied on in making this determination must be retained in the files. If the generator bases its determination on testing results, it must retain all waste-analysis data.

8.16   REPORTING AND RECORD-KEEPING OBLIGATIONS

The federal RCRA program requires that each large-quantity generator that ships
its hazardous waste off-site prepare and submit to EPA, no later than March 1 of
each even-numbered year, a single copy of the Biennial Report on EPA Forms
8700-13 and 8700-13A (EPA, § 262.41). As with other elements of the RCRA
program, state report and record-keeping rules might differ from EPA requirements, so generators should consult with their state agency for specific requirements. Some states require more frequent "Periodic" reporting. The purpose of a
Biennial (or Periodic) Report is to inform the EPA or state agency of the type and
quantity of hazardous waste shipped and the identities of the transporters and
TSD facilities used during the preceding period. Such reports must also discuss
waste-minimization efforts.

In accordance with the federal RCRA program, all generators must retain the
following records for the periods specified:

- a copy of each signed manifest form and TSD receipt copy for three years
  from the date the waste was accepted by the initial transporter
- a copy of each notice, certification, and determination pertaining to the
  land-disposal restriction requirements for at least five years after the waste
  to which they pertain was sent off site
- a copy of each Biennial Report and Exception Report for a period of at
  least three years from the due date of the report
- records of any test results or waste analyses made in accordance with a
  hazardous-waste determination for at least three years from the date that
  the waste was last sent for treatment, storage, or disposal
- a copy of the generator's contingency plan or spill prevention and control
  plan
- a copy of the generator's training records for current personnel must be
  retained until the facility closes, and for former employees such records
  must be retained for at least three years from the date the employee last
  worked at the facility (EPA, § 262.40).

When a file is created to retain records to meet the above requirements, it should
be labeled with the following legend: "These records are subject to mandatory
retention periods. Do not destroy without first obtaining the approval of the
responsible employee." The records subject to the mandatory record-retention

periods must be made available to authorized EPA or state-agency inspectors upon their request. Applicable state requirements can be more stringent, and generators should consult with their state agencies.

8.17   MARKING FOR OFF-SITE TRANSPORT

Generators must take the following three steps before sending hazardous waste off-site:

- Generators must make certain that all hazardous waste containers ready for transportation have been properly sealed, marked, and labeled (EPA, § 262.30, § 262.31, § 262.32).
- Generators must prepare and route a manifest (EPA, § 262, Subpart B).
- Generators must make certain that the transporter's vehicle is properly placarded for the hazardous waste shipment (EPA, § 262.33).

The placard DANGER should appear in highly visible fashion on each side and each end of the vehicle. The vehicle should also bear, on each side and each end, a placard that relates to the hazardous trait of the waste being shipped, for example, FLAMMABLE or POISON. Placards should be:

- printed in English
- readily visible from the direction the vehicle faces
- located away from any marking that would reduce their effectiveness
- located so that dirt and water will not be directed to them from the wheels of the motor vehicle
- maintained on the vehicle in such a way as to prevent obscurement and deterioration

Prior to their shipment for off-site management, containers of hazardous waste must adhere to rules imposed by both EPA and the Department of Transportation (DOT). EPA rules require that every container of hazardous waste being shipped off-site bear the following marking:

- HAZARDOUS WASTE—Federal Law Prohibits Improper Disposal. If found, contact the nearest police or public-safety authority or the U.S. Environmental Protection Agency.
- Generator's name and address
- Manifest document number

This marking must be:

- durable, in English, and printed on or affixed to the surface of the container
- displayed on a background of sharply contrasting color
- unobscured by other labels
- located away from any other markings that could reduce its effectiveness (EPA, § 262.32)

8.18   PERSONNEL TRAINING AND EMERGENCY PROCEDURES

Environmental statutes impose emergency preparedness and spill-response obligations, such as mitigation and reporting, on industrial facilities based on the nature and quantity of the chemicals or wastes stored on-site. In addition, the Occupational Safety and Health Administration (OSHA) has adopted its Hazardous Waste Operations and Emergency Response Rule which prescribes associate training, safety precautions, and procedures that must be followed in responding to emergencies involving hazardous substances (EPA, § 1910.120). Thus, there is substantial overlap between and among environmental emergency-response requirements.

Each generator must train employees how to respond to emergencies, including:

- procedures for using, inspecting, repairing, and replacing facility emergency and monitoring equipment
- communications or alarm systems
- response to fires or explosions
- response to groundwater contamination
- potential shutdowns (EPA, § 262.34(a)(4), EPA, § 265.16)

Each generator must keep on file (1) a written description of the type and amount of training given to employees, (2) the job title, and the name of the employee holding the job, for each position involving hazardous-waste handling, and (3) job descriptions for these job titles, including the requisite degree of training to hold the job.

EPA requires generators to foreclose the possibility of fire, explosion, or accidental hazardous-waste release through:

- a communications (public-address) system capable of instructing personnel in the event of an emergency
- procedures for summoning the local fire department or emergency-response team in the event of an emergency
- fire extinguishers and other appropriate fire-control equipment, spill-control equipment, and decontamination equipment
- sufficient water volume and pressure to operate fire-control equipment (EPA, § 262.34(a)(4), EPA, § 265 Subpart C)

Aisles must be kept free from obstruction to allow fire, spill-control, or emergency-response teams access to any area where such teams might need access. Generators should have arrangements with local authorities for responding to emergencies.

Generators must have a contingency plan designed to minimize hazards to human health or the environment in the event of fires, explosions, or accidental release of hazardous waste (EPA, CFR § 262.34(a)(4), 265 Subpart D). A copy of the contingency plan must be maintained. The following is a checklist for evaluating a contingency plan:

a. provides for at least one employee (the emergency coordinator) either on the premises or on call (i.e., available to respond to an emergency by reaching the facility within a short period of time) with the responsibility for coordinating all emergency-response measures

b. ensures that the location of fire extinguishers, spill-control material, and fire alarm and the telephone number of the fire department, unless the facility has a direct alarm, and is readily available to the emergency coordinator

c. includes a plan for responding to releases of hazardous substances (which include hazardous wastes)

d. minimizes hazards to human health or the environment from fires, explosions, or any unplanned releases of hazardous waste or waste constituents

e. describes the actions that employees who are responsible for emergency response must take with respect to an unplanned sudden or nonsudden release

f. describes arrangements with local emergency-response teams

g. lists names, addresses, and phone numbers of all persons qualified to act as emergency coordinator

h. has been updated to reflect current operations

i. includes a list of all emergency equipment, including the location and a physical description of each item on the list and brief outline of its capabilities

    j. identifies an evacuation plan for facility personnel

    k. ensures that designated employees successfully complete a program of training (taught by a person trained in hazardous-waste-management procedures and including instruction relevant to the positions in which they are employed) that teaches them to perform their duties in a way that ensures compliance with the requirements and an effective response to emergencies

8.19 POLLUTION-PREVENTION PLANS

Some states require facilities that generate over a certain amount of hazardous waste to adopt and implement pollution-prevention plans. Other states require such plans both as a result of generating a certain amount of hazardous waste and of exceeding certain thresholds of toxic chemical releases. Generators should consult with their state agencies to identify pollution-prevention plan requirements.

8.20 UNDERGROUND TANKS

A widely recognized environmental concern in recent years is soil and groundwater contamination caused by underground storage tanks (USTs). Several factors contribute to UST contamination, including corrosion of tanks and piping, structural failure, or poor work practices resulting in spills when tanks are emptied or overflows when tanks are filled. EPA rules govern existing and new tanks (EPA, Federal Register 1988).

For existing tanks, the goal of the regulatory plan is to determine whether the tanks should be brought up to a certain design or operating standard, or whether they should be closed. Further, UST owners and operators are responsible for a variety of items including: the maintenance and upkeep of the tanks to prevent leaks, financial responsibility requirements, and any corrective actions that may result from a defective tank. For example, if a tank was determined to have a leak present which was causing an environmental problem, the UST program would require the owner or operator of the tank to take corrective action to eliminate the environmental problem whether by changing the design or closing the tank completely.

For new tanks, the law has been designed such that strict design and operating standards are required. Additionally, while there is no federal permit required for a UST, it is the owner's and/or operator's responsibility to notify the government, through an elaborate notification system, that an UST is being installed. State laws vary widely and regulatory managers should study them closely.

8.21 ABOVEGROUND STORAGE TANKS

In addition to RCRA requirements, OSHA imposes requirements for storing flammable or combustible chemicals and other hazardous substances in aboveground tanks (OSHA CFR 1910.106). State health departments or fire departments may regulate storage tanks. The National Fire Prevention Association (NFPA) has rules for tanks containing flammable and combustible liquids—diesel, gasoline, and oil. Aboveground tanks are regulated indirectly under the Clean Water Act (USC, § 1251–1387) and the Oil Pollution Act (USC, § 2701–2761). Clean Water Act requirements are discussed in Chapter 6. A facility is exempt from the Oil Pollution Act if it has aboveground storage capacity of 1,320 gallons (4,900 liters) or less, and no single tank has a capacity over 660 gallons (2,490 liters). Otherwise, when a tank is sited:

- A spill-prevention, -control, and countermeasure plan is required if a discharge of oil in harmful quantities could affect navigable waters (EPA, § 112).
- The facility must be inspected and the plans certified by a registered professional engineer. The plans must meet good engineering practices.
- The plans must have management approval and provide for site security.
- Plans must include a prediction of the direction, flow rate, and quantity of oil that could be discharged.
- Plans must provide for appropriate containment or diversion to prevent the spill from reaching the water.
- Copies of the plans must be kept at the facility or the nearest field office and updated every three years.

Institutions should examine the adequacy of a spill plan during any acquisition of property that has aboveground storage-tank capacity. Like aboveground tanks, underground tanks that hold petroleum products are regulated under the Clean Water Act. The Oil Pollution Act requirements discussed above are triggered for underground tanks with a capacity of over 42,000 gallons (158,760 liters) (EPA, § 112.1).

8.22   MEDICAL-WASTE INCINERATION AND DISPOSAL

Wider use of disposable personal protective equipment and a higher level of awareness to medical hygiene as a result of AIDS has led to an increased amount of medical waste. In the past, most medical waste was sent to landfills untreated. Highly infectious material from microbiology departments tended to be auto-claved. Autoclaving may not be acceptable for this purpose if the temperature of the waste, not only the temperature in the drain, is not high enough to penetrate the core load and kill all organisms present (New Directions in Federal Regulations 1991). Many hospitals have incinerators. Now, many states require that regulated medical waste be rendered noninfectious and destroyed.

State and local governments have assumed the principal role in regulating medical waste. These programs are based on the federal 1988 Medical Waste Tracking Act (MWTA) which established a five-state pilot program that expired in 1991 (USC, § 6992-6992k). Under MWTA, generators in these states who transported more than 50 pounds (22.6 kilograms) of medical waste per month off-site had to initiate a tracking system similar to the manifest required under RCRA (EPA, § 259.52, § 259.74, § 259.75, § 259.81, § 259.90, § 259.91). Forty-three states have regulations controlling medical or infectious waste (Menicucci and Coon 1994). These laws differ in the types and scope of medical waste they regulate.

Most states recommend incineration of medical waste. There are more than 6,000 hospital-waste incinerators nationally. Because medical waste is largely unregulated at the federal level, and most of the waste burned by hospitals and universities is ordinary trash and not hazardous waste, an RCRA treatment, storage, and disposal permit is not required. Although these facilities are not deemed to be treating, storing, or disposing of hazardous waste, the ash they generate must be properly disposed if found to be hazardous.

At present, medical-waste incinerators are estimated to be the largest source of U.S. dioxin emissions, generating about 95% of the dioxin in the air (EPA 1994). Because of this, hospitals can face local opposition when constructing their own incinerators and the trend is to ship infectious waste away for commercial treatment.

An EPA-proposed standard for medical-waste incinerators under the Clean Air Act is expected to cut dioxin emissions from this source by 99%. The standard is controversial because it is forcing technological advances in incinerator design that cannot be met by many existing hospital incinerators. The standard will also reduce emissions of nine other harmful constituents by 93%, or 88,000 tons (89,400 metric tons) annually: lead, carbon monoxide, mercury, particulate matter, sulfur dioxide, nitrogen dioxide, cadmium, and hydrogen chloride. As a result of this standard, more universities and hospitals will need to make arrangements to ship medical waste off-site for incineration and disposal, or install upgraded incineration systems.

States lack authority to regulate the interstate transport of medical waste. In a 1992 report, EPA suggested that a uniform federal program would be necessary to address interstate transport (U.S. EPA 1990a). Congress has yet to enact new federal authority to regulate medical waste. However, 1988 amendments to the Clean Water Act prohibit the discharge of medical waste into navigable waters (USC, § 1311 (f)). This law carries a potential for $1 million in fines and 36 years' imprisonment.

Congress has also restricted ocean-disposal of medical waste within 50 nautical miles of land through the U.S. Public Vessel Medical Waste Anti-Dumping Act of 1988 (USC, § 2503). The Ocean Dumping Ban Act of 1988 prohibits permits for disposing medical waste in the territorial seas of the United States (USC, § 1402(k), § 1402(a)).

Some states and localities have tried to enact laws restricting the disposal of medical waste that was generated outside of their boundaries (*BFI Medical Waste v Whatcom County* 1993). A Washington state county ordinance was invalidated but a Baltimore, Maryland, ordinance has been upheld. In general, however, states and localities cannot prohibit waste from coming in or going out of their area (*Philadelphia v New Jersey* 1978, *Fort Gratiot v Michigan Dept. of Natural Resources* 1992, *Chemical Waste Mgmt. v Hunt* 1992, *Oregon Waste Systems v Dept. of Environmental Quality* 1994, *C & A Carbone v Town of Clarkstown* 1994) and Congress is unlikely to restrict the flow of waste across state lines anytime in the near future. However, adequate supplies of waste has been a problem for local disposal facilities whose financing is contingent upon receiving and processing a guaranteed amount of waste. Thus, Congress may enact "flow-control" legislation in the near future that will allow states and localities to manage their own wastes by requiring that waste within their boundaries be disposed of at designated facilities.

There have been convictions for mail fraud, conspiracy, and racketeering based on medical-waste mismanagement (*U.S. v Paccione* 1990). In one case, a doctor was convicted for submitting false information to a state to obtain medical-waste transport and disposal permits to illegally dispose of the waste. In another case, a doctor was convicted for improper disposal of blood and other medical waste into the Hudson River (*U.S. v Villegas* 1991). The laboratory also was indicted and its participation in the state's Medicaid program was suspended (*U.S. v Plaza Health* 1993). Doctors and hospitals are also potentially liable for civil and criminal fines for the illegal activities of those with whom they contract to handle disposal, even if the disposal company represents itself as duly licensed (*U.S. v Paccione* 1990).

Because of the lack of a uniform federal standard for regulating medical-waste disposal, there is a significant degree of duplication and complexity in managing medical waste. EPA air regulations mean that institutions will need to rely more on large regional incinerator facilities to dispose of medical and laboratory waste. In general, a smaller range of waste materials is entering landfills. Many states are banning from landfills items such as lead-acid batteries, yard waste, unprocessed tires, combustible materials such as waste paint and solvents, used oil, and large appliances (Nat'l Solid Waste Mgmt Assn. 1990). Liquid wastes are prohibited from being disposed of in municipal landfills. Materials are being removed and separated for recycling or composting. Special waste landfills are being established for incinerator ash, construction debris, and other specific waste streams (Solid Waste Management 1992).

8.23 SPECIFIC WASTE STREAMS OF POTENTIAL CONCERN

The safety manager of the institution cannot be an expert in all the chemicals that might be used; such a goal is impossible to achieve. This set of specific items may serve as an "early warning" that if one of these waste streams is known to exist, the facility might have a hazardous-waste obligation deserving special attention.

Agricultural Waste

Pesticide waste from the use and application of pesticides at agricultural facilities at land-grant colleges across the country may need to be managed and disposed of under RCRA hazardous waste requirements. Cow manure is considered a non-hazardous waste, although its storage may be regulated by EPA with regard to non–point source pollution requirements under the Clean Water Act.

Arts and Theater Waste

Art and theater departments use chemicals in printing, textiles, ceramics, painting, silk screening, sculpture, and photography. Paints, thinners, heavy metals, waste ink, and solvents are commonly found in these departments. These are discussed in greater detail in Chapter 13. Photography also generates waste silver and rinsing and developing solutions. These leftover chemicals may trigger hazardous-waste management requirements. For example, hazardous wastes may be generated in the use of inorganic acids, zinc, acetic acid, diesel, kerosene, turpentine, alcohols, oils, and oil-based paints.

Ash

Hospitals burn infectious materials, surgical instruments, rubber, and plastic in medical incinerators. Incineration concentrates toxic substances in these materials in the ash residue. This complicates ash disposal. Ash may leach significant quantities of metals, such as lead, cadmium, mercury, and arsenic. A 1989 *Newsday* article reported that ash residue from New York's Lutheran Medical Center's medical-waste incinerator had high levels of dioxin, dibenzofurans, lead, cadmium, chromium, zinc, and barium (Newsday 1989).

The cost of disposing of hazardous waste once skyrocketed, but is now decreasing as a general trend because of substantial capacity that was built in response to local waste-management planning requirements under RCRA. Nevertheless, new Clean Air Act emission requirements for medical-waste incinerators may force hospitals to shift a large amount of medical-waste disposal to outside disposal options. Thus, medical facilities should have a plan for acceptable waste-analysis and waste-handling practices. For those who continue to operate their own incinerators, proper incinerator waste-testing and disposal practices will continue to be needed. Incinerator ash must be separated from other waste and disposed of properly.

Batteries

The generation, transport, and collection of spent batteries is not subject to RCRA regulation. Facilities that store spent batteries before reclaiming are regulated by EPA under RCRA (EPA, § 266).

Biotechnology Waste

Biotechnology research sources include laboratory counting procedures, residue from research and spent reagents, cleaning of laboratory equipment, and lead contaminated during the process, and backflush of resin filters and changeouts (EPA/530-SW-90-057, 7). Organic solvents and other organic chemicals are typical of this waste stream.

Building and Construction Waste

Twenty percent of public and commercial buildings in the United States have asbestos-containing building fixtures (U.S. EPA 1988). By now, the health risk associated with asbestos and its relationship to lung disease (e.g., asbestosis) when released and exposed in friable (dust) form is well recognized. The environmental professional is well aware of the need to check for the presence of asbestos in building ceiling and floor tiles, wallboard, duct work, insulation materials, and high-temperature-area materials prior to performing any renovation or demolition activities (Menicucci and Coon 1994, 542).

EPA regulates asbestos dust emissions during renovation or demolition of institutional, public, or residential (five or more dwelling units) buildings under the Clean Air Act (USC, § 7412, EPA, § 60.140-60.157). Owners and operators subject to these requirements includes an institution that owns, leases, operates, controls, or supervises the facility being demolished or renovated or controls or supervises the company doing the work (EPA, § 61.141, § 61.145). A licensed contractor must be used. The asbestos-containing material (ACM) must be specially transported and disposed. OSHA regulations also apply.

Asbestos is classified as a hazardous substance under CERCLA, so that institutional property owners may be liable for past disposal. Courts have held that the sale of a building for demolition constituted arranging for disposal of the asbestos in the building under CERCLA (*C.P. Holdings* v *Goldberg-Zoino* 1991)

Most construction/demolition debris is disposed of in landfills. The requirements for disposal are not as stringent as the requirements for disposing of other types of solid-waste since its contents are relatively inert. However, some construction debris contains hazardous materials. Besides asbestos, lead pipes, PCBs, and toxics in paints and treated lumber may be found in construction waste. These hazardous materials must be removed and handled separately. While difficult to separate, recyclable materials in construction waste includes bricks, asphalt, and cinder blocks.

Chemotherapy

The hospital or medical center that treats cancer patients with aggressive chemical therapeutic agents must dispose of the waste from the chemicals. Some cytotoxic agents used for chemotherapy are listed as hazardous wastes, including: cyclophosphamide (U058), daunomycin (U059), melphalan (U150), mitomycin C (U010), streptozoticin (U206), and uracil mustard (U237) (Menicucci and Coon 1994).

Fluorescent Bulbs

Fluorescent bulbs are present in and around institutional buildings in a variety of places—in classrooms, operating rooms, sickrooms, hallways, and outside lighting. The mercury content of fluorescent, high-pressure sodium, mercury vapor, and metal halide lamps may cause the lamps to be subject to hazardous waste regulation under RCRA (EPA Federal Register 1994).

EPA has proposed two options for regulating mercury-containing lamps. The first option will exempt the lamps from regulation as hazardous waste under RCRA. The second option would exempt mercury-containing lamps from certain handling practices, but would still subject the lamps to "special-waste" disposal requirements.

As a whole, EPA estimates that over 300 million lamps and up to 64,000 facilities could be affected by the proposal. The Agency estimates industry would save almost $100 million per year if the lamps are exempt from RCRA hazardous-waste regulations altogether. The "special-waste" option would result in a significantly smaller savings of $18 million per year.

If the "special-waste" option is chosen, however, end users of mercury-containing lamps in their curing operations would be required, among other burdensome provisions, to label boxes and drums, notify EPA of their status as a hazardous-waste generator, transport the lamps via a hazardous-waste transporter, and manage the lamps consistent with land-disposal restrictions.

The proposed changes should make it easier to dispose of spent lamps, though the degree of relief will depend on the proposed regulatory option EPA selects, and on the way in which individual states administer their municipal solid-waste and hazardous waste programs. In the meantime, EPA assumes that spent fluorescent lights will fail RCRA's Subtitle C toxicity characteristic (TC) test (i.e., the lamps will contain mercury in concentrations of 0.2 mg/L or more). The Agency's "debris" variance, which had allowed generators of spent mercury-containing lights to dispose of them as ordinary solid waste, ended in May of 1995. Thus, unless a generator can demonstrate that its spent lighting passes the TC test, it must handle them as a Subtitle C waste, at least until EPA finalizes its current proposal. This means that generators need to subject a representative sample to RCRA's test and, if results are positive, meet all the disposal requirements of RCRA Subtitle C, including the use of licensed hazardous waste transporters and the hazardous-waste manifest system.

Alternatives to handling mercury-containing lamps as other than hazardous waste are few at the state level. Although recycling of mercury-containing waste is an attractive option, and many states are actively exploring this alternative, only a handful (Pennsylvania, California, and Minnesota) have mercury-recycling programs in operation at this time. New Jersey and Texas are in the process of starting up similar programs. For more information on the subject, an EPA Publication ("Lighting Waste Disposal"), is available through the Agency's "Green Lights Customer Service Center" at (202) 775-6650 (phone) or (202) 775-6680 (fax).

Laboratory Waste

Laboratory wastes are typically generated in quantities of less than one gallon (3.78 liters) per occurrence (U.S. EPA 1990b). Many labs produce only small amounts of hazardous waste. In educational institutions, chemistry labs tend to be the largest hazardous waste generators (U.S. EPA 1990b). Organic chemistry waste sources include inorganic salts, benzene, and halogenated and nonhalogentated solvents. Inorganic chemistry generates iodine, xylene, naptha, p-dichlorobenzene, mercury, chromium, lead, carbon tetrachloride, and heavy-metal waste. Another source of large quantities of surplus reagents and waste chemicals left in laboratories occurs when researchers leave or retire from the institution.

Biology departments use organic solvents, acids, alkalies, carcinogenic compounds (benzidine, diaside, and nitrogenous compounds used in assays), and mercury salts. Molecular biology may use: phenol, methanol, formaldehyde or formalin (for the preservation of specimens), ethers, acrylonitrides, organic solvents, heavy metals, inorganic acid washes, and inorganic salts (U.S. EPA 1990b).

Geology, physics, psychology, and engineering generally use fewer chemicals and generate less waste than chemistry and biology. Chemicals used in these departments commonly include organic solvents, acids, bases, and metals.

Material waste in a laboratory may not be a hazardous waste if an RCRA-listed hazardous chemical is not the sole active ingredient in the concentration or formulation (EPA, § 261.33). Environmental professionals or attorneys should be consulted to determine whether the formulation is hazardous.

Practices such as disposal in lab packs have been developed to minimize exposure to hazardous substances. Lab packs are handled in accordance with U.S. Department of Transportation (DOT) requirements. Each type of waste is collected in a discrete container and labeled. These small bottles, vials, and cans are separated by compatibility of contents, and packaged in drums with absorbent cushioning (e.g., vermiculite) sufficient to protect against breakage (New Directions in Federal Regulations 1991). The drums are labeled, sealed, and shipped to authorized disposal or incineration facilities. A lab-pack record must accompany each barrel. Lab-pack disposal combines small packages. This saves space in a lab, in contrast to the use of separate barrels for each kind of waste.

While not the primary target of RCRA, high-school and clinical laboratories are not exempt from federal regulations and state scrutiny. For example, Illinois has drafted legislation, currently in the Illinois General Assembly, that would require the Illinois Environmental Protection Agency to develop a program to collect and dispose of school waste. The legislation comes in response to two 1993 incidents. The first occured when hazardous material collected in a load of garbage exploded in the parking lot at the middle of a Peoria school, injuring four people. Several weeks later, a complaint by a concerned parent led to the subsequent removal of boxes of chemicals from another school's boiler room (Eckert 1995).

Maintenance Waste

Maintenance shops perform a wide variety of functions, including electrical, painting, plumbing, hardware, signs, grounds and building care, sheet metal, plant operations, custodial, and fleet services. Maintenance operations, including machine and motor-vehicle maintenance, generate waste oils, solvents, pesticides, water-treatment chemicals, PCB-containing fluids from old transformers, possibly asbestos, and other kinds of wastes.

A particular issue arises in maintenance work that involves the replacement of transformers and other equipment. The dielectric fluid in electric equipment, transformers, capacitators, and hydraulic systems may contain polychlorinated biphenyls (PCBs), particularly in equipment manufactured through the 1970s. EPA regulates PCB spills, and the use, disposal, and phaseout of PCB-containing equipment in concentrations of 50 parts per million (ppm) or more under the Toxic Substances Control Act (USC, § 2601-2671, EPA, § 761 et. seq.). EPA regulations require disposal of regulated equipment and PCB-containing fluid in an EPA-approved method, such as in a chemical-waste landfill, an authorized incinerator, or a high-efficiency boiler. Certificates of destruction or disposal should be obtained from the disposal facility. Spills of 50 ppm or more concentration are also considered disposal and must be handled accordingly. State waste oil or "special-waste" regulations may apply to PCB concentrations of less than 50 ppm (EPA, § 761.60(d)).

Worn-out or broken industrial appliances such as stoves, refrigerators, washing machines, and clothes dryers can be sold to scrap processors who use shredders to recover metal components for reuse to produce new steel. Many scrap-metal dealers require that PCB-containing components be removed before the appliances are processed.

Large institutions are likely to have transformers on their property. PCBs are classified as a hazardous substance under CERCLA, so institutional property owners may be liable for releases. The primary health concern of PCBs is associated carcinogenic effects and conversion to dioxin during burning. Owners of transformers should inspect them regularly for fluid leaks. Most transformers can be drained of PCB-containing fluid and replaced with non-PCB fluid. Drained PCB fluid should be disposed of in the manner specified above.

Medical Waste

Health-care institutions produce a great volume of medical waste, including large quantities of infectious waste (O'Reilly 1992). Medical waste is 0.3 percent of the 158 million tons of solid waste generated annually (Medical Waste Specialists 1991), and hospitals generate 77% of medical waste nationwide. Some experts feel that if regulations were based on epidemiological and microbiological information, the only two types of medical waste that would be regulated are sharps (syringes) and microbiological waste (Rutala 1991). Unfortunately, much more debris is regulated, due largely to public reaction when infectious waste began washing ashore in 1988. Regulated medical wastes may include several or all of the following: cultures and stocks of infectious agents and associated biologicals; pathological waste; human blood, blood bags, and blood products; used needles, syringes, and other sharps; bandages and vials; contaminated animal carcasses; surgery or autopsy waste; dialysis waste; discarded medical equipment; isolation waste; various plastics; and disposable personal protective clothing.

A significant amount of confusion exists because a uniform federal standard is lacking for regulating medical-waste disposal. State regulations must be consulted because they may use different definitions of "regulated medical waste" and may impose different requirements. Based on state estimates, up to 50–60 pounds (27 kilograms) of blood-related waste (including bed materials, towels, disposable personal protective clothing, sponges and sharps) can be generated by one open-heart surgery (*Chicago Tribune* 1989). A 400-bed hospital can generate over 1,300 pounds (590 kilograms) on a daily basis. While infectious waste may be just 5% to 7% of a hospital's total waste stream, the cost of disposal is comparably expensive, varying from 10 cents to $1.50 a pound (*Chicago Tribune* 1994). OSHA regulations govern worker exposure, handling, and marking before transport. Those requirements are addressed in other chapters.

Radioactive Waste

Medical and academic institutions generate low-level radiological waste in research, teaching, and medical treatment. Radiopharmaceuticals such as iodine-125 (fat absorption and kidney function), gallium-67 (tumor imaging), and fluorine-18 (bone imaging), hydrogen-3, carbon-14, and other materials used in *in vitro* testing, cancer treatment, and cancer research are some examples (Menicucci and Coon, 1994, 550–551). Plastic, cloth protective covers, and glass can become contaminated during experiments or medical treatment.

Treatment, handling, and packaging requirements for the radioactive components of mixed waste require a knowledge of the radionuclide identities and concentrations in the waste as well as its physical form, the radioactive waste class (Class A, B, or C), and the chemical form. Nuclear Regulatory Commission (NRC) regulations specify minimum waste form and stability requirements for radioactive waste (EPA, § 61). Many of these materials have short half-lives (such as ^{125}I, ^{32}P, and ^{35}S). Temporary storage may allow decay so that they need not be managed as radioactive waste. Also, some states may allow small quantities of these materials to be disposed of in authorized municipal solid-waste landfills, so long as they are not mixed with hazardous waste (Texas Admin. Code 1993–94).

The environmental professional should have a radiation safety program in place and know at any given time the location and quantity of the radioactive material that is located on campus or at a hospital facility. A central, computerized inventory of radioactive substances is recommended, as is an active role by faculty or staff in the administration and policing of radiation safety policies. The NRC conducts routine audits and a university or hospital may be cited for inventory accountability problems (*Columbus Dispatch* 1994).

RCRA-regulated hazardous waste that also contains low-level radioactive waste is referred to as "mixed waste." Medical and academic institutions generate mixed waste in diagnostics and treatment, at pilot plants, reactors, and through laboratory research. In 1990, academic institutions generated 28,982 cubic feet and medical facilities generated 19,904 cubic feet of mixed waste (EPA/530-SW-90-057, 47). The large majority of this mixed waste is liquid scintillation fluid wastes. One major university can produce 980 cubic feet of low-level mixed waste per year (*Columbus Dispatch* 1994). Mixed-waste-generating practices include the following (EPA/530-SW-90-057, 7):

- laboratory counting procedures
- residue from research and spent reagents
- cleaning of laboratory equipment
- cleaning of contaminated components
- lead contaminated during processes

The hazardous-waste component typically includes organic solvents of liquid-scintillation cocktails, organic chemicals, metallic lead, hazardous-waste oil, phenol or chloroform, chromate and cadmium wastes, and corrosive liquids (EPA/530-SW-90-057, 6-7).

For example, mixed waste with metal is generated through decontamination of lead used as shielding, from batteries, paint wastes, and lead-containing research solutions (Oak Ridge National Laboratory 1993). Mercury-contaminated equipment and debris, and mercury from laboratory experiments, are also sources of metal-containing mixed waste. Organic chemicals found in mixed waste include chloroform, trichloroethane, methylene chloride, waste oils, CFCs, and other chlorinated organics used in research or as pesticides. Aqueous corrosive mixed wastes, primarily acids, are also generated from laboratory operations. Other sources of mixed waste include biological wastes, incinerator ash, and filter bags.

For mixed waste, the hazardous-waste component is regulated by EPA and the radioactive waste is regulated by the Nuclear Regulatory Commission (NRC) under the Atomic Energy Act (AEA). Under EPA land-disposal restriction regulations, mixed waste must be treated to an applicable treatment standard for the hazardous portion of the waste before land-disposal is permitted or a variance is granted (EPA, § 268 Subpart D). EPA treatment standards are either expressed as concentration levels or treatment technologies. Metal-containing mixed waste requires a diverse set of treatments. The recommended treatment for most liquid-scintillation fluids mixed with organics (organohalides, chlorinated organics, fluorinated organics) is by incineration. EPA's selected treatment for aqueous corrosives is neutralization. Incineration is also an option. EPA has three treatability groups for low-level radioactive mixed waste that cannot simply be treated with the best demonstrated available technologies (BDAT) for the hazardous component involved (EPA/530-SW-90-057, 11):

- *Macroencapsulation:* the BDAT standard for radioactive lead solids, which entails applying surface coating materials to reduce surface exposure in the event of leaching
- *Amalgamation:* the BDAT for radioactive elemental mercury. This requires combining with reagents such as copper, zinc, nickel, gold, and sulfur and results in a semisolid amalgam that reduces potential mercury vapor emissions
- *Incineration:* the BDAT standard for radioactive hydraulic oil contaminated with mercury

The environmental professional should be knowledgeable about the identity of treatment facilities in their area and their acceptance criteria. If adequate treatment capacity is not available, the institution must store the mixed waste in accordance with RCRA until the capacity is available or seek an EPA variance.

The legal landscape for mixed-waste generators is complex because of the array of regulations and their interpretation by individual states. State requirements should also be consulted for the regulation of mixed waste, because NRC may delegate its authority to the states. The Low-Level Radioactive Waste Policy Act of 1980 resulted in 43 states organizing themselves into nine compacts primarily to consolidate their disposal for this waste. Other states, the District of Columbia, and Puerto Rico have their own rules for responsibly disposing of low-level waste and mixed waste. However, there are few low-level radioactive waste–disposal sites, and the closure in July 1994 of a Barnwell, South Carolina, disposal site means that hospitals and universities will have to stockpile their waste until an alternative site is available. Department of Transportation packaging and transportation requirements also apply.

Tires

Used tires generated by universities with their own fleet operations, or by school districts with school bus fleets, may be well suited to recycling or reuse. Rather than stockpiling used tires, the environmental professional should consider potential recycling options that may be available in their area. These include sale for retreading or recapping, reuse in smaller rubber parts such as rubber mats, tire-derived fuel, use in rubberized paving materials, or for use in playground equipment or reef construction.

Used Oil

Pouring used motor oil down storm drains, onto the ground, or into the trash can contaminate groundwater, surface water, and soils, and can bring penalties and bad publicity for the institution. At least 25 organic chemicals that are listed hazardous wastes may be found in significant concentrations in used oil (EPA/530-SW-90-057, 6). It only takes one gallon (3.785 liters) of oil to ruin one million gallons of water.

EPA has decided not to list used oil as a hazardous waste. However, used oil may still be declared a hazardous waste if it exhibits a toxic characteristic. Under EPA regulations, waste oil destined for disposal and not exhibiting a hazardous characteristic can be incinerated without a RCRA permit. Nevertheless, states have the right to impose additional controls, and some have designated used oil as a hazardous waste. Environmental professionals should consult state and local regulations to determine the regulatory status of waste oil.

As of 1988, over half the states either had or were planning to start used oil-recycling programs. It is important that used oil not be mixed with other substances such as gasoline or paint thinner when contributing used oil for recycling. Mixing can contaminate the oil and make it unfit for recycling.

8.24 WASTE MINIMIZATION

Apart from regulatory or social incentives, a waste-minimization program can reduce disposal costs, administrative overhead, and potential liabilities. Some general waste minimization techniques appear in Table 8–3 (U.S. EPA 1990a).

Designing pollution out of current methods and practices begins, for medical and academic institutions, right in the lab. Waste-minimization options may start with a centralized purchasing program that would allow for:

- sharing of chemicals among common users
- ordering chemicals in exact amounts to be used and shipments on an as-needed basis. Net savings from unit-cost purchasing can be easily overwhelmed by eventual disposal costs if the chemicals are not used
- first-in, first-out chemical stock usage (American Chemical Society 1985)
- training for procurement staff to emphasize life-cycle costs, particularly the exit costs of unused chemicals

The amounts of chemicals required for laboratory analyses have been reduced as technological advances increase. Compared to wet chemistry techniques, one-

Table 8–3. Waste-Minimization Methods for Research and Educational Institutions

Category	Waste-Minimization Methods
Improved Material-Management Practices	Establish a centralized purchasing program.
	Order reagent chemicals in exact amounts.
	Encourage chemical suppliers to become responsible partners (e.g., accept outdated supplies).
	Establish an inventory-control program that can trace chemical from cradle to grave.
	Rotate chemical stock.
	Develop a running inventory of unused chemicals for use by other departments.
	Centralize waste management. Appoint a safety/waste-management officer for each department.
	Educate staff on the benefits of waste minimization.
	Perform routine self-audits.
Improved Laboratory Practices	Scale down the volumes of chemicals used in laboratory experiments.
	Increase use of instrumentation.
	Reduce or eliminate the use of highly toxic chemicals in laboratory experiments.
	Pre-weigh chemicals for undergraduate use.
	Reuse/recycle spent solvents.
	Recover metal from catalyst.
	Treat or destroy hazardous-waste products as the last step in experiments.
	Keep individual hazardous-waste streams segregated, segregate hazardous waste from nonhazardous waste, segregate recyclable waste from nonrecyclable waste.
	Assure that the identity of all chemicals and wastes is clearly marked on all containers.
	Investigate mercury recovery and recycling with an outside vendor.
Improved Practices in Other Departments	Replace oil-based paints with water-based paints in art instruction and maintenance operations.
	Modify paint-spraying techniques.
	Reduce generation of pesticide waste.
	Collect waste oil and solvents for recycling.
	Use biodegradable aqueous or detergent cleaners.
	Investigate silver recovery or recycling with an outside vendor for photoprocessing wastes.
	Provide training in hazardous-waste management practices for students in art and photography courses and facilities management/maintenance personnel.

tenth to one-hundredth of the volume of a substance is needed for instrumental analysis (U.S. EPA 1990a). Common instruments that reduce chemical usage include nuclear magnetic-resonance analysis, chromatography (gas, liquid, thin layer), mass spectrophotometry, atomic absorption, photoionization detectors, ion probes, X-ray diffraction analyzers, infrared (IR) and ultraviolet (UV) spectrophotometers, and magnetic balances (U.S. EPA 1990a). However, the use of wet chemistry techniques may be an essential part of the education process.

In addition to increased use of instrumentation, other practices that can reduce laboratory waste generation include:

- scaling down the volumes of chemicals used in laboratory experiments
- substituting less-hazardous chemicals
- pre-weighing chemicals for group experiments
- decreasing reagent solvent use through recycling and reuse of solvents in cleaning
- increasing the amount of in-lab destruction of chemical wastes as a final step in experiments (National Research Council 1983)
- establishing a central area for waste storage, segregation, and treatment
- because waste contaminated with a hazardous substance may become hazardous waste, keep individual waste streams and recyclable waste segregated
- clearly identifying chemicals and wastes and marking all containers

In addition to achieving waste minimization, such practices decrease fire and explosion hazards, and reduce harmful organic vapors in laboratory air.

One way to organize a waste-minimization program is to address institution-wide goals first, and then individual department goals. Individual department minimization efforts may also be encouraged through practical application of the "polluter pays" principle. This is an economic incentive to reduce waste and can be done by adopting accounting practices that apportion waste-management costs to the departments that generate the waste. This could lead users of extremely hazardous chemicals and high-volume users to develop strategies that reduce these costs in their department budgets.

For grounds maintenance, the use of irrigation injection may replace spraying operations to reduce the amount of pesticide waste and rinse waters generated. Another alternative is to apply pesticides in dry powder form and water.

Figure 8.1 contains a number of waste-generation worksheets developed by EPA in conjunction with EPA's Waste Minimization Assessment Manual. These may be helpful in assessing your waste-minimization program.

8.25   SUPERFUND: THE COMPREHENSIVE
ENVIRONMENTAL RESPONSE, COMPENSATION,
AND LIABILITY ACT AND SITE CLEANUP

The Comprehensive Environmental Response, Compensation, and Liability Act (CERCLA) is a complicated law that has been extensively assessed in other sources (O'Reilly, Kane, and Norman 1993). The law allocates cleanup liability among past contributors of waste to a site, and it provides financing for cleanups through a Hazardous Substance Response Trust Fund which is largely funded by industry through taxes on crude oil and certain chemical feedstocks (USC § 9611). Financial liability for cleanups is primarily placed on "potentially responsible parties" (PRPs), those entities who are responsible for the contamination of the particular site, for example, a college whose chemistry-lab waste went into a landfill along with mixed municipal and residential wastes.

When a waste site is found, studied, and listed for action, the federal EPA can and does conduct initial emergency-response actions at the site. EPA then supervises long-term site cleanups. To recover costs of the response and remediation actions, EPA sues the companies that it asserts were PRPs, seeking an order for reimbursement of EPA's expenditures and overhead costs, typically in the millions of dollars. Emergency response can include such actions as soil removal and incineration and provision of alternative drinking-water supplies and relocation of communities. Long-term cleanups are called "remedial actions," and involve soil and groundwater treatment and monitoring, as well as public participation in the process (USC § 9617, § 9621). Private-property acquisition is also permitted. A well-known example of EPA exercising its Superfund authority to the broadest extent, including the acquisition of property by the government, was at the hazardous-waste site known as "Love Canal" in Niagara Falls, New York, in the 1970s. That large-scale and controversial federal action gave rise to many of the extensive powers vested in EPA by the 1980 Superfund law and its later 1986 amendments (USC § 9661).

Compliance with hazardous-waste-disposal requirements does not preclude liability under CERCLA (Wiener and Bell 1994). Liability for the cost of these cleanups extends to the owner and operator of a facility, any person who arranged for disposal or treatment, and any person who accepted hazardous substances for transport (USC § 9607). EPA can seek reimbursement for the entire cost of a cleanup against any PRP, and the PRP can then seek legal redress and proportional contribution from other PRPs. Provisions are made in the law for *de minimis* settlements with PRPs that are responsible for only a minor portion of the response costs (USC § 9622(g), Wiener and Bell 1994).

Sites owned, operated, or otherwise controlled by a hospital or academic institution can be "facilities" subject to Superfund cleanups. In addition, they may be held liable for the costs of a Superfund action at a site they do not own or control. Those defined as a "person" who may be liable in this manner include an individual, corporation, commercial entity, a state, municipality, or political subdivision of the state (USC, § 9601). Thus, a university, a nonprofit or proprietary hospital, and a local school district are "persons" subject to this law. They may be held liable as a PRP if they arranged for treatment or disposal at an off-site facility (*Alcan Aluminum* v *Cornell University* 1993).

CERCLA faces almost certain overhaul by Congress. It is heavily criticized as accomplishing too little at great expense. The extent of site cleanup that should be achieved, the unlimited, retroactive liability that CERCLA places on PRPs, and the federal government's role in managing the cleanup process are issues that Congress will reexamine in the near future.

8.26   HAZARDOUS-SUBSTANCE-RELEASE REPORTING

The Superfund (CERCLA) also established an ongoing notification requirement for reporting unpermitted release to the environment of certain hazardous substances to EPA (USC § 9601). The type of release that needs to be reported are spills, leaking, and similar emergency situations. A period of 24 hours should be used for measuring the released quantity.

The notification applies to any person in charge of a facility from which the hazardous substance has been released to the environment in a reportable quantity (O'Reilly, Kane, and Norman 1993). "Person in charge" is not defined under CERCLA, but presumably is more broad than "owners or operators" and includes universities, hospitals, and schools.

A list of CERCLA hazardous substances that must be reported is located in EPA's regulations (EPA, § 302). EPA has established "reportable quantities" (RQ) for these substances, and releases above the RQ must be reported to EPA's National Response Center (1-800-424-8802) (EPA, § 302.4). EPA requires the owner or operator of a facility from which a hazardous substance is released to provide notice to the affected community as well (USC, § 9611).

For example, suppose there was an unpermitted release of 1,500 pounds of styrene from your facility. By referring to Table 302.4 in the regulations, the RQ for styrene is 1,000 pounds (453 kiliograms) and you would need to report. Other hazardous substances besides those listed need to be reported as well. For RCRA wastes that exhibit the characteristic of toxicity, the RQ is based on the contaminant on which the characteristic of toxicity is based. If a hazardous substance does not have a CERCLA RQ, and is not hazardous based on the characteristic of toxicity, a RQ of 100 pounds (45.3 kiliograms) applies.

REFERENCES

Alcan Aluminum v *Cornell University,* 990 F.2d 711 (1993) (contribution obtained from Cornell under a CERCLA action).

American Chemical Society. 1985.

BFI Medical Waste Systems Inc. v *Whatcom County,* 983 F.2d 991 (9th Cir. 1993).

C&A Carbone v *Town of Clarkstown,* 114 S. Ct. 1677 (1994).

Chemical Waste Management v *Hunt,* 504 U.S. 334 (1992).

Chicago Tribune. 1989. Revisions urged in Medical Waste Law. 16 May, page 1.

Chicago Tribune. 1994. Seeking Gold in Medical Waste; Opportunities, Risks Run High for Disposal Firms. 30 January, page 1.

10 Code of Federal Regulations Part 61.

29 Code of Federal Regulations § 1910.120.

29 Code of Federal Regulations Part 1910, Subpart H.

40 Code of Federal Regulations §§ 60.140–61.157.

40 Code of Federal Regulations § 112 (1994).

40 Code of Federal Regulations § 112.1.

40 Code of Federal Regulations §§ 259.52, 259.74–75, .81, .90–91 (1990).

40 Code of Federal Regulations § 260.10.

40 Code of Federal Regulations § 261, Appendix II.

40 Code of Federal Regulations § 261.3(a) (2).

40 Code of Federal Regulations § 261.5.

40 Code of Federal Regulations § 261.5(a).

40 Code of Federal Regulations § 261.21.

40 Code of Federal Regulations § 261.22.

40 Code of Federal Regulations § 261.24.

40 Code of Federal Regulations § 261.31.

40 Code of Federal Regulations § 261.32.

40 Code of Federal Regulations § 261.33.

40 Code of Federal Regulations Part 262, Subpart B.

40 Code of Federal Regulations § 262.12.

40 Code of Federal Regulations § 262.20(a).

40 Code of Federal Regulations § 262.20(e).

40 Code of Federal Regulations § 262.23.

40 Code of Federal Regulations § 262.30.

40 Code of Federal Regulations § 262.31.

40 Code of Federal Regulations § 262.32.

40 Code of Federal Regulations § 262.33.

40 Code of Federal Regulations § 262.34(a).

40 Code of Federal Regulations § 262.34(a)(1)(i).

40 Code of Federal Regulations § 262.34(a)(1)(ii).

40 Code of Federal Regulations § 262.34(a)(4).
40 Code of Federal Regulations § 262.34(c).
40 Code of Federal Regulations § 262.34(d).
40 Code of Federal Regulations § 262.34(e).
40 Code of Federal Regulations § 262.34(f).
40 Code of Federal Regulations § 262.40.
40 Code of Federal Regulations § 262.41.
40 Code of Federal Regulations § 262.42(a).
40 Code of Federal Regulations § 262.42(b).
40 Code of Federal Regulations Part 265, Subpart C.
40 Code of Federal Regulations Part 265, Subpart D.
40 Code of Federal Regulations Part 265, Subpart I.
40 Code of Federal Regulations Part 265, Subpart J (except § 265.197(c) and § 265.200).
40 Code of Federal Regulations § 265.16.
40 Code of Federal Regulations Part 266.
40 Code of Federal Regulations Part 268.
40 Code of Federal Regulations § 268 Subpart D.
40 Code of Federal Regulations § 761 et seq.
40 Code of Federal Regulations § 761.60(d).
Columbus Ohio Dispatch. 1994. OSU Careless with Radiation, the NRC Says. 28 May, page 1D.
C.P. Holdings, Inc. v *Goldberg-Zoino & Associates, Inc.,* 769 F. Supp. 432 (D.N.H. 1991).
Eckert, Toby. 1995. Legislation for Disposal of Chemical Waste at Schools Clears Illinois Senate. *Environmental News* 4 (April).
EPA. 1994. Draft Dioxin Reassessment Report.
EPA. 1988. 53 Federal Register 37082 (Sept. 23, 1988).
EPA. 1993. 58 Federal Register 31114 (May 28, 1993).
EPA. 1994. 59 Federal Register 38288 (July 27, 1994).
EPA. 1994. Lighting Waste Disposal. (EPA 420-R-94-004) (March 1994).
Fort Gratiot Sanitary Landfill v *Michigan Department of Natural Resources,* 504 U.S. 353 (1992).
Martin v *Kansas Board of Regents,* 1991 Westlaw 33602 (D. Kan).
Medical Waste Associates Ltd. Partnership v *Mayor and City Council of Baltimore,* 966 F.2d 148 (4th Cir. 1992).
Medical Waste Specialists Loom as Biotech Ally. *Boston Business Journal* 2, no. 1 (25 February): 13.
Menicucci, Margaret, and Cheryl Coon. 1994. Environmental Regulation of Health Care Facilities: A Prescription for Compliance. *Southern Methodist University Law Review* 47: 537–577.
National Research Council. 1983. Prudent Practices for Disposal of Chemicals from Laboratories.

New Directions in Federal Regulations on Hazardous Waste. 1991. *Medical Laboratory Observer* 23 (April): 31.

National Solid Waste Management Association. 1990. Recycling in the States. *National Solid Waste Management Association: Mid-Year Update.* Page C-5.

Newsday. 1989. Hospital Toxin Levels Tested High; But No Evidence of Incinerator Violations at Lutheran. 19 February.

Oak Ridge National Laboratory. 1993. National Profile on Commercially Generated Low-Level Radioactive Mixed Waste. (NUREG/CR-5938 ORNL-6731).

Oregon Waste Systems v *Department of Environmental Quality of Oregon,* 114 S. Ct. 1345 (1994).

O'Reilly, James. 1992. *State & Local Government Solid Waste Management.* Deerfield IL: Clark Boardman Callaghan.

O'Reilly, James, Kyle Kane, and Mark Norman. 1993. *RCRA and Superfund Practice Guide.* 2d ed. Colorado Springs:Shepards/McGraw-Hill.

OSHA, Occupational Safety & Health Administration, 29 Code of Federal Regulations § 1910.106.

Philadelphia v *New Jersey,* 437 U.S. 617 (1978).

Rutala, William, and David Weber. Infectious Waste: Mismatch between Science and Policy. 1991. *New England Journal of Medicine* 325 (22 August): 578.

Solid Waste Management Practices and Developments, Part 2. 1992. *Public Works* 123 (August):9: 73.

25 Texas Administrative Code. Sec. 289.11–126 (West Supp. 1993–94).

United States v *Paccione,* 751 F. Supp. 368 (S.D.N.Y. 1990).

United States v *Plaza Health Labs,* 3 F.3d 643 (2d Cir. 1993) and see prior decisions, *Plaza Health Labs* v *Perales,* 702 F. Supp. 86 (S.D.N.Y.), aff'd, 878 F.2d 577 (2d Cir. 1989).

United States v *Villegas,* 784 F. Supp. 6 (S.D.N.Y. 1991).

15 United States Code § 2601–2671.

33 United States Code § 1251–1387.

33 United States Code § 1402(k).

33 United States Code § 1412(a).

33 United States Code § 2503.

33 United States Code § 2701–2761.

42 United States Code § 302.

42 United States Code § 6903(27).

42 United States Code § 6941–6949a.

42 United States Code § 6992k.

42 United States Code § 7412.

42 United States Code § 9601.

42 United States Code § 9601 et. seq.

42 United States Code § 9607.

42 United States Code § 9611.

42 United States Code § 9617.

42 United States Code § 9621.

42 United States Code § 9622(g).

42 United States Code § 9661.

U.S. EPA. 1990a. Guide to Pollution Prevention: Research and Educational Institutions, EPA/625/7-90/010 (June 1990).

U.S. EPA. 1990b. Low Level Mixed Waste, A RCRA Perspective for NRC Licensees. EPA/530-SW-90-057 (Aug. 1990), page 6.

U.S. EPA. 1990c. Medical Waste Management in the United States, Second Interim Report to Congress PB91-130187.

U.S. EPA. 1988. Study of Asbestos-Containing Materials in Public Buildings; A Report to Congress (Feb. 1988).

Wiener, Robin, and Christopher Bell. 1994. *RCRA Compliance and Enforcement Manual.* 2d ed. Colorado Springs: Shepards/McGraw-Hill.

Chapter *9*

Worker Training and Safety Issues

9.1 INTRODUCTION

The Occupational Safety and Health Act (OSHA) applies to private-sector employers. OSHA was passed in 1970 in the midst of a flurry of social and environmental legislation prompted by the perception that government action was necessary to correct what were considered to be failures of the market system in the United States—to "engineer" social change through legislation. The lofty goal of the Act was stated in section 651(b): "to assure so far as possible every working man and woman in the nation safe and healthful working conditions and to preserve our human resources." In 1971, there were 4.2 million establishments and over 57 million employees covered by the Act. By 1994, the private-sector workplace had grown to more than six million workplaces and over 95 million employees. Contrary to common perception, coverage extends to all those workplaces, even those with fewer than ten employees.

For nearly the entire history of OSHA, Congress has in effect exempted employers with ten or fewer employees from so-called programmed enforcement inspections, subjecting them to inspection only in cases of complaints or accident investigations. These same employers are exempt from certain record-keeping requirements by regulation. But small employers can be cited for the same kinds of violations of substantive standards as large employers, including such wide-ranging standards as the Confined Spaces, Lockout/Tagout, and Hazard Communication Standards that are found at 29 C.F.R. §§1910.146, .147, and .1200, respectively.

Originally, Congress exempted federal, state, and local employees from these violations. Protection for executive-branch employees was provided by the adoption of Executive Order 12196 in February 1980 under which all federal agencies were ordered to develop health and safety programs and to comply with OSHA standards. Early in 1995, as part of the House Republicans' Contract with America, Congress extended the same requirements to its own operations in Public Law 104-1, finally providing protection under the OSHA for employees of Congress.

State and local employees are covered in all the 23 states (and two territories) with state plans presently approved by OSHA, including several that provide for monetary penalties where the agencies fail to comply with state standards. At the time of the original statute's passage, Congress recognized the states' interest in workplace health and safety and devised a system under which the federal agency would set minimum workplace standards, allowing the states, if they so chose, to play a significant role. Direct coverage of state and local government employees was not included in the original act because of the belief that Congress had no power to regulate the conditions of employment between the state governments and their employees. Some believe that Supreme Court decisions over the last 20 years have modified that conclusion.

Although Congress provided that OSHA coverage should not extend to industries regulated by other comprehensive schemes, the agencies responsible have largely been displaced in practice if not in actuality in many of these industries. Railroads, trucking, certain operations of the marine industry, and others now find themselves facing OSHA inspections. Industries where the exemption is still viable include mining, where a comparable but far more intrusive regulatory scheme is in place; sea-based operations under Coast Guard jurisdiction; certain aspects of rail operations which the Federal Railroad Administration (FRA) is in charge of; and aspects of the nuclear-energy industry. The trend among the agencies, as illustrated by the recent case involving the FRA, is to defer to OSHA's expertise except in the areas where the industry's characteristics presents unique hazards and the Agency has unique expertise.

9.2   THE OCCUPATIONAL SAFETY AND HEALTH REVIEW COMMISSION

When Congress established OSHA, a major controversy with the legislation was the question of how appeals to the civil citations and penalties would be adjudicated. The final compromise led to the establishment of the Occupational Safety and Health Review Commission (OSHRC), an independent agency made up of three members and a first level system of administrative law judges (ALJs). The

Review Commission's findings of fact are binding, but as to issues of interpretation of the standards, the Secretary of Labor holds sway. In a 1991 decision (*Martin* v *OSHRC* 1991), the Supreme Court said that OSHA's greater expertise and experience in enforcement and standards-setting require courts to defer to OSHA when there is a dispute over the meaning of a standard. OSHA has taken the position in recent cases that the Commission has little or no role to play in deciding the meaning of language in the standards. In an era when OSHA is expanding its reach by writing standards in general terms with ambiguous and unclear language in the name of writing "performance-oriented" standards—albeit at the behest of industry—the need for a neutral third party to review the Agency's pronouncements is even more critical. For the moment, however, the Agency's memoranda and policy statements, which are not published in the Federal Register, are essential keys to understanding an employer's obligations.

9.3   THE NATIONAL INSTITUTE FOR OCCUPATIONAL SAFETY AND HEALTH

Congress also recognized the need for an organization to focus attention and resources on causes of occupational injuries and illnesses. Toward that end, Congress created the National Institute for Occupational Safety and Health (NIOSH) in the former Department of Health, Education, and Welfare. Born out of the Bureau of Occupational Safety and Health, NIOSH is located within the Public Health Service's Centers for Disease Control. Its tasks include, in addition to its research function, providing support for the Educational Resource Centers, conducting health-hazard evaluations at the request of employers, and providing supporting documentation to OSHA in the standards-setting process.

NIOSH has largely abandoned its efforts to survey the scientific literature and to provide summaries on health effects of chemicals to OSHA and the regulated community. OSHA has been left to its own devices to develop the scientific justification for its decisions. In part, this may have been the result of the perception that NIOSH was unconnected to the real world, in that its recommendations were rendered without regard to issues of technical or economic feasibility. This attitude is perhaps best illustrated by its policy when dealing with cancer-causing substances. NIOSH takes the policy position that, because a safe level cannot be established with present-day scientific knowledge, exposures must be eliminated or reduced to the lowest detectable level. While this may be an admirable position for an idealist to take, it has left the impression that NIOSH has not kept up with scientific progress and the widespread recognition that risks related to cancer and other health effects vary for equal exposures to different chemicals.

9.4   ROLE OF THE STATES

Congress recognized that many states had effective programs of worker protection. As with air, water, and hazardous-waste regulatory plans, Congress included provisions in OSHA permitting states (1) to exercise jurisdiction over any occupational safety and health issue where no federal standard exists; (2) to assume responsibility for development and enforcement of state rules equivalent to federal standards; (3) to operate an occupational health and safety program comparable in funding, authority, and staffing to the federal program; and (4) to enforce more stringent standards where they are compelled by local conditions and do not unduly burden interstate commerce.

Section 18 requires a state to submit a plan to OSHA if the state desires to assume responsibility for the development and enforcement of standards "relating to" any occupational safety or health issue with respect to which a federal standard has been issued. Congress provided for funding of up to 50% of the cost of administering and enforcing programs in the states, and up to 90% to improve state capabilities in this area.

Twenty-three states and two territories (see Table 9–1) have state plans approved by the Secretary of Labor under this provision of the Act.

Although the law provides for withdrawal of approval by the Secretary after notice and hearing, no state plans have had their approval involuntarily withdrawn. In the early 1990s, as a result of allegations of ineffective operation of the state's plan, OSHA proposed to withdraw approval of North Carolina's pro-

Table 9–1.   States with OSHA "State Plans"

Alaska	Michigan	Tennessee
Arizona	Minnesota	Utah
California	Nevada	Vermont
Connecticut*	New Mexico	Virgin Islands
Hawaii	New York*	Virginia
Indiana	North Carolina	Washington
Iowa	Oregon	Wyoming
Kentucky	Puerto Rico	
Maryland	South Carolina	

*Covers state employees only

gram. The state vigorously opposed OSHA's action, and succeeded in retaining jurisdiction after making significant changes in funding, resources, and enforcement.

The states with approved state plans promulgate and enforce their own standards. Generally, the states have six months after promulgation of a new federal standard to adopt an equivalent standard under their own rules. The majority of states simply adopt the federal standard. In some cases, particular states have acted in advance of the federal government because of the perception that certain needs were not being addressed; in several cases, the states have adopted standards that are significantly different from the federal rule. For example, Maryland has had provisions limiting exposure to lead in the construction industry for more than ten years, while the federal standard was adopted only after Congress imposed a statutory deadline. California has adopted a standard that addresses chemical process safety issues, that applies to a smaller group of industries than does the federal rule, while covering a larger list of chemicals at lower quantities.

Approximately ten of the states with state plans have adopted or retained state standards that are more restrictive than the federal rules, or which address areas in which there are no comparable federal standards. Many of these standards were carried over from state health and safety programs that existed prior to the passage of OSHA. Some examples include standards on elevators, boiler and pressure vessels, confined spaces, and exterior window cleaning on skyscrapers. A number of the states have more expansive hazard-communication standards as well.

9.5   RECORD-KEEPING AND OTHER OSHA STANDARDS

The broad scope of OSHA's standards is shown in Table 4–3, the index of OSHA standards. The chemical-specific standards often include requirements for the development of specific exposure and medical records, administrative memoranda, reports, and certifications. The lead standard, for example, requires that employers address a large number and variety of specific requirements (shown in Section 4.16), each of which requires or implies that the employer maintain some form of record. The record keeping is, of course, an enormous burden, but more importantly, creates a significant problem for assuring compliance. Not only do the records have to be kept, they must be maintained. This is interpreted by OSHA to mean that the information in the records must be updated when it changes.

Some of the OSHA standards which imply significant record-keeping requirements do so because they require some kind of periodic inspection. For example, the standard on *Portable Wood Ladders* requires that

Ladders shall be inspected frequently and those which have developed defects shall be withdrawn from service. . . .

This provision does not explicitly require records. However, most employers
would find records useful if an inspector found an employee using a ladder that
had a defect.

As this illustrates, the OSHA ladders rule could be interpreted to require that
at least four types of records be kept to demonstrate employer compliance with
the inspection requirement. Of course, not all employers will have such records,
and in fact, many do not keep them. Nevertheless, because the burden is on the
employer of demonstrating compliance with the standard when the prima facie
case is made in the Agency's complaint, records of such kinds of activities are always desirable.

Table 4–5 is a listing, by no means complete, of other OSHA standards that
have either written program, training, or inspection requirements. Nearly every
standard that is proposed by OSHA today has some provision for additional
records. Most OSHA proposals also have significant record-keeping provisions.
We can anticipate that all future standards will be equally designed to provide an
adequate record for inspectors to use in evaluating compliance.

9.6 RETENTION OF AND ACCESS TO RECORDS

OSHA does not have a generic record-keeping standard mandating the kinds of
records to be kept. There is a single provision that addresses retention periods and
authorized access to medical records. The retention provisions require employers
to maintain medical and exposure records for 30 years, and provide for the transfer of records to the National Institute for Occupational Safety and Health
(NIOSH) in the event the employee intends to dispose of them or ceases to do
business. Medical records are defined to exempt health-insurance claims records,
first-aid records of one-time treatment and subsequent observation of cases not
involving medical treatment, and records of employees employed for less than
one year who are given the records on termination of employment. The exemption for first-aid records applies essentially to those cases that would otherwise
not be recordable on the OSHA Form 200.

Paragraph (e) of this section establishes the right of employees and their
designated representatives to obtain access to and copies of medical and exposure
records. Employers may not charge for initial copies of records, and must provide
access within 15 working days of a request. Employee medical records are subject to a provision that can limit access by the employee if, in the opinion of an
employer's physician, there is information regarding a terminal illness or psychi-

atric condition that could be detrimental to the employee's health. In such cases, the employer may provide the information to another physician of the employee's choosing after denying the employee access in writing to the detrimental information. Confidential information identifying persons who have provided information about the employee may be excised from the record provided to the employee.

As OSHA compliance becomes more complex, increasing numbers of standards require the completion and maintenance of records. The challenge for employers today is to assure that the records are being kept and that they are accurate, and it is by no means an easy task.

9.7   OSHA ILLNESS AND INJURY RECORD-KEEPING

The injury and illness record-keeping regulations promulgated by the Department of Labor impose three basic obligations on employers. *First,* employers must maintain a log and summary (the OSHA Form 200) of recordable occupational injuries and illnesses for each business establishment. OSHA regulations define "establishment" as a "single physical location where business is conducted or where services or industrial operations are performed." However, if distinctly separate activities—for example, different employers are co-located at a single site— are performed at a single physical location, each activity should be considered a separate establishment and OSHA records should be maintained separately.

Second, employers must prepare a supplementary, detailed record, the OSHA Form No. 101, *Supplementary Record of Occupational Injuries and Illnesses,* for each recordable occupational injury or illness. The employer may keep the required information on other forms, such as workers' compensation forms, Supervisor's First Report of Accident forms, or other insurance forms, so long as all the necessary information is present. Finally, for each establishment employers must post in the month of February a signed copy of the prior year's annual summary (Form 200), which describes that facility's recordable occupational injuries and illnesses for the past calendar year. Occasionally, an employer may also be requested to complete a statistical survey form containing occupational injuries and illnesses and return it to the Bureau of Labor Statistics.

For each recordable occupational injury or illness, the log must contain a separate entry which identifies the employee and briefly describes the injury or illness. The employer is required to record this information for recordable injuries and illnesses as soon as possible, but no later than six working days after learning that a recordable occupational injury or illness has occurred.

OSHA requires that employers must record an injury or illness and keep the record current for the full five-year period that the employer is required to retain its log. For example, an employer may discover some time after the fact that a former employee was injured or became ill due to a work-related event while still employed by the employer. If this discovery is made within the five-year record retention period, OSHA contends that the employer must record that injury or illness on the log for the year in which the case occurred, even though the involved employee may have not worked at the facility for several years.

Third, employers must update the log of illnesses and injuries when new information about outstanding cases is obtained. Information on the log must be current within six days of the employer finding out about the case. If the log for a remote location is maintained at a central office, such as in the case of retail store chains, the log at the work site must be current within 45 days. Thus, employers who have aggressive back-to-work programs must update the log to reflect the total days away from work for temporarily disabled employees each time the employee is contacted as part of the return-to-work program.

The Bureau of Labor Statistics (BLS) annual survey of employers is used to develop the statistical data reported in the annual survey of occupational injuries and illnesses in the United States by industry. These data are collected once a year by mail from approximately 20,000 selected employers. Reports for each year's data are reported by the BLS in aggregate form, and individual employer records of injuries and illnesses supplied to BLS are not made available to OSHA.

When there is a change in ownership in an establishment, OSHA interprets the retention requirement to relieve the new owner of the obligation to update the records. The regulations require the new owner to preserve the records of the prior owner, and to record cases only for that portion in the year in which he owns the establishment.

Posting the Log

An employer must post in each of its establishments a copy of the summary of the previous calendar year's recordable occupational injuries and illnesses for the establishment by February 1 of the following year. The summary must be presented on the right-hand portion of the OSHA No. 200 form. The summary is designed to remove personal identifiers and the specific description of the illness or injury. It must include the calendar year covered, the company's name, the facility's address, a signature certifying that the information contained on the annual report is correct, the title of the person signing, and the date of the certification. This summary must be posted throughout the month of February in a conspicuous place or places where notices to employees are customarily posted. If an employer has employees who do not regularly report to work at a single facility, the employer must mail or present a copy of the summary to those employees at the appointed time. Note that the person signing the summary is specifically subject to the Section 17(g) sanctions under § 1904.5 (d)(2).

As mentioned above, the *Log and Summary* (OSHA No. 200), the *Supplementary Record* (OSHA No. 101), and the annual summary must be retained by the employer for five years following the end of the year to which they relate. Upon request, employees, former employees, and their representatives must be provided access to the records for examination and copying in a reasonable manner and at reasonable times. The regulations require that mandated records be provided, upon request, for the purpose of inspection or copying, to OSHA inspectors as well as to representatives of the Secretary of the Department of Health and Human Services. Former employees are specifically included among those who are to be provided access to the records kept pursuant to § 1904. It has been held that the disclosure of these records does not violate the Fourth-Amendment right against unreasonable search and seizure because it authorized the access without a warrant. When Form 200 is posted, whatever reasonable expectation of privacy the employer might have had in the information is lost, and lacking this element, the access requirement under the rule did not violate the Constitution.

Despite an ongoing legal debate over an employer's rights to withhold required data in the absence of a warrant or subpoena, many employers have long made such information available for review by OSHA inspectors. Depending on the circumstances, firms often conclude that it is more beneficial to be cooperative than confrontational, reasoning that an evaluation of injury and illness records might limit the scope of an inspection. This judgment may be based in part on the relative ease with which the health and safety agency might obtain a warrant or subpoena. Nevertheless, OSHA citations are occasionally vacated when challenged on the grounds of employer objections to the lack of a warrant, and the employer's right to require a warrant is firmly established even for the most unobtrusive request by OSHA inspectors.

Illness and Injury Record-Keeping Exemptions

The standard generally applies to all employers to whom the OSH Act applies. However, the regulations provide exceptions for small employers and employers classified in certain industries. Employers who have fewer than 11 employees at all times during a calendar year are required only to log all cases, to report fatalities and multiple hospitalizations, and to complete the BLS statistical survey if requested. Small employers do not have to complete the OSHA No. 101 form for each occupational injury or illness, nor do they have to compile and post an annual summary of recordable injuries and illnesses.

OSHA exempts employers in certain industries because they are considered low-hazard industries. Generally, the exempted industries include employers in certain retail trades, financial services, insurance, real estate, and other service-based industries. For employers whose establishments are mixed use, the predominant classification by the largest share of production, sales, or revenue controls the classification of the establishment, although payroll or employment may be used where the primary economic activity is not accurately measured by revenue measures. The exempted industries include most but not all classifications in Standard Industrial Classification Codes 52–89.

Recordable Cases

As suggested by the statutory language described above, the following injuries and illnesses come within OSHA's record-keeping obligations:

1. occupational fatalities, regardless of the time between the injury and death or the length of the illness
2. occupational injuries and illnesses, other than fatalities, that result in lost workdays
3. nonfatal occupational injuries or illnesses that do not involve lost workdays but result in the transfer of the employee to another job or termination of employment, require medical treatment (other than first aid), or involve loss of consciousness or restriction of work or motion. Also included in this last category are diagnosed occupational illnesses that are reported to the employer (but are properly not classified as fatalities or lost workday cases).

To help an employer decide whether a particular case needs to be recorded, the BLS suggests the following five-step analysis:

1. Determine whether a case occurred (that is, whether there was a death, injury, or illness).
2. Determine whether the case was work-related (whether it was caused, contributed to, or was aggravated by an event or exposure in the work environment).
3. Decide whether the case is an injury or an illness.
4. If the case is an illness, record it and check the appropriate illness category on the log.
5. If the case is an injury, decide if it is recordable based on a finding of medical treatment, loss of consciousness, restriction of work or motion, or transfer to another job.

The regulation defines nonfatal cases to include "any *diagnosed* occupational illnesses. . . ." "Illnesses" are defined on the back of Form 200: An "occupational illness" is any abnormal condition or disorder (other than one resulting from an occupational injury) caused by exposure to environmental factors associated with employment. It includes acute and chronic illnesses or diseases that may be caused by inhalation, absorption, ingestion, or direct contact.

The BLS guidelines also identify hearing loss as a recordable illness if it is determined to be work-related. OSHA currently cites employers for failure to record STS exceeding 25 dB when calculated according to the procedure specified in the current noise standard.

Determination of Work-Relatedness

As a general rule, injuries that occur *anywhere* on the employer's premises are considered work-related. Thus, an injury that occurs in a company bathroom or a hallway is an occupational injury. The major exceptions are injuries that occur on company recreational facilities and in company parking lots, which are not recordable unless the employees were engaged in some work-related activity. It is immaterial for record-keeping purposes whether the employer or employee was at fault for the injury or illness or whether the injury or illness was preventable.

OSHA concluded that the presumption of work-relatedness is necessary when cases occur on the employer's premises to keep the recording criteria simple. Nevertheless, the presumption is rebuttable and the case is not recordable if there is no relationship between the person's presence on the premises and his status as an employee.

Examples used in the BLS guidelines referring to horseplay and employees choking while eating lunch in a company cafeteria are considered recordable by OSHA. But employers must determine whether a case is work-related in other more obscure situations, such as when a condition manifests itself at the job site but is unrelated to the individual's status as an employee.

Reporting of Fatality or Multiple Hospitalization Accidents

If a workplace accident causes the death of an employee or the hospitalization of three or more employees, the employer must report the accident orally or in person within eight hours of becoming aware of the circumstances. Reports may be made to the nearest OSHA office or to the national reporting hot line, 1-800-321-OSHA.

Retention of and Access to Exposure and Medical Records

The principal generic OSHA record-keeping standard, 29 C.F.R. 1910.20, addresses retention periods for medical and exposure records and authorized access to medical records. Unless specified by a specific standard, employers must maintain medical and exposure records for 30 years, and provide for the transfer of records to the National Institute for Occupational Safety and Health (NIOSH) in the event the employer intends to dispose of them or ceases to do business. Medical records are defined to exempt health-insurance-claims records if they are maintained separately and not accessible by the employer by employee name or other personal identifier; records prepared solely for litigation; and records of voluntary employee-assistance programs. Medical records that do not need to be retained include health-insurance-claims records, first-aid records of one-time treatment and subsequent observation of cases not involving medical treatment, and records of employees employed for less than one year who are given the records on termination of employment. The exemption for first-aid records applies essentially to those cases that would otherwise not be recordable on the OSHA Form 200.

Employee-exposure records must also be kept. Data on exposure monitoring must be kept, although background information, such as worksheets and laboratory reports, need only be kept for one year, provided other data relevant to the sampling and analytical methodology are retained for the full period of 30 years. Material-safety data sheets (MSDS) do not need to be retained if some other record of the chemical identity, the location where it was used, and the time period of use are kept. Biological monitoring results, such as blood-lead tests, are considered exposure records.

Any analysis of medical or exposure records must also be kept for 30 years. The records can be kept in any form, manner, or process provided the information can be reliably retrieved. Chest X-rays must be kept in their original state.

The designated representative of the employee includes the legal representative of a deceased employee. Former employees, including those who are actively involved in union-organizing campaigns, are also entitled to access these records.

A separate provision authorizes OSHA access to the records as authorized by the statute. Regulations found at 29 C.F.R. 1913.10 define agency practice and procedure for gaining access. The rules, issued to satisfy privacy concerns, limit the types of requests that can be made, the agency personnel who may be granted authority to access the records, and the uses to which the records are put.

The provisions in paragraph (f) of § 1910.20 protect trade secrets from disclosure except in cases of medical emergency or for purposes of assessment of listed occupational health needs. These include hazard assessment, exposure measurement, selection of personal protective equipment, and related studies for design of engineering controls. In such cases, the person seeking the data may be required to agree to keep the information confidential and to sign confidentiality agreements. Under extraordinary circumstances, the Assistant Secretary of Labor may impose additional conditions on release of trade-secret information.

Finally, employees must be told of the existence, location, and availability of records subject to these provisions, the name of the person responsible for their maintenance and for providing access, and the employee's rights to access the records. As is often the case with OSHA standards, a copy of the standard must be made available to employees and kept by the employer.

9.8 EMERGENCY-RESPONSE TRAINING

Employers who have hazardous chemicals in their workplaces must establish either (1) an emergency-response plan according to the provisions of 29 C.F.R. § 1910.120(q), or (2) an emergency-action plan according to § 1910.38(a). An employer may only choose not to have an emergency-response plan if employees are required to evacuate an area when an emergency occurs and the employees are not permitted to assist in handling the emergency. Hazardous chemical spills are considered emergency releases if the spill is of such a size that it cannot be cleaned up by employees working in the area or department, or by maintenance employees.

Paragraph (q) of § 1910.120 defines the elements of an emergency-response plan and specifies the training required according to the anticipated duties of the employee. Employees who simply evacuate are not covered. Employees who are likely to observe an emergency release and are to notify the proper authorities in that event must receive training defined as "first responders awareness level." Employees who respond by attempting to protect persons or property nearby, but who are not expected to try to stop the release, are trained as the "first responder operations level." A hazardous-materials technician is a person who participates as an active member of the response team, and the hazardous-materials specialist is one who can act as site liaison with government officials. Finally, the on-scene incident commander must be trained in all aspects of emergency response and must have specialized knowledge about the emergency-response plan, hazards of the particular chemicals on site, and how to coordinate with the local emergency-response team and federal regional response teams.

Training for each of these different levels must be provided by a person who has demonstrated (1) the technical knowledge of the subject matter they are to teach, and (2) academic credentials or instructional experience necessary to demonstrate competent instructional skills.

Employers who choose not to have an emergency-response team must prepare and train employees in the emergency-action plan. The plan must include emergency escape procedures and route assignments; procedures for employees to follow to operate critical operations before they evacuate; procedures to account for all employees upon evacuation; rescue and medical duties for those employees who are to perform them; the preferred means of reporting emergencies; and the names or job titles of persons who can be contacted for further information about the plan. An alarm system, distinct from other purposes, must be used. Employees who will assist in the orderly and safe evacuation must be trained when the plan is put into place and the plan must be reviewed with all other employees when it is developed and each time employees' duties change or the plan changes. For employers with fewer than ten employees, the plan need not be written, but may be communicated orally to the employees.

9.9   HAZARD-COMMUNICATION REQUIREMENTS

In November 1983 the Occupational Safety and Health Administration (OSHA) of the U.S. Department of Labor issued its final Hazard Communication Standard (HCS); a standard for laboratory safety was adopted in 1990. It is designed to "reduce the incidence of chemically related occupational illnesses and injuries among employees in the manufacturing sector." The Standard was developed on the assumption that the wider availability of workplace hazard information required by the standard will allow employers to implement more effective safety measures and provide workers with the information they need to protect themselves.

To ensure that the hazardous properties of chemicals are evaluated and communicated to employers and employees, the HCS requires manufacturers and importers to evaluate the hazards of substances they manufacture or import, to develop material-safety data sheets (MSDSs) and warning labels, and requires all employers to conduct employee training and maintain warning labels and MSDS for hazardous chemicals.

Summary of the HCS

The Hazard Communication Standard establishes uniform hazard-communication requirements for manufacturers. Each employee working in a manufacturing facility who is exposed or may potentially be exposed to hazardous chemicals in the workplace must receive information and training that is appropriate for the potential exposures involved. Chemical manufacturers and importers are required to evaluate the hazards of the chemicals that they produce or import, and pass this information to downstream employers through the use of labels on containers and material-safety data sheets (MSDS). Each employer's hazard-communication program must provide the necessary hazard-chemical information to its employees (through container labels, MSDS, and training) so that they can understand and implement the protective measures developed for their workplaces.

Exemptions

OSHA's Hazard Communication Standard contains a number of significant exemptions which are incorporated into the Company's program. Certain additional items are *completely exempt* from the provisions of the OSHA Standard. These include:

1. hazardous waste as defined in the Solid Waste Disposal Act, as amended by the Resource Conservation and Recovery Act, when it is subject to regulations issued under that statute by the Environmental Protection Agency
2. tobacco or tobacco products
3. wood or wood products
4. articles
5. foods, drugs, or cosmetics intended for personal consumption by employees while in the workplace

The *labeling requirements* of the OSHA Standard are not applicable to certain substances:

1. pesticides, as defined in the Federal Insecticide, Fungicide, and Rodenticide Act, when subject to the labeling requirements of the Act and implementing regulations issued by the Environmental protection Agency
2. foods, food additives, color additives, drugs, and cosmetics (including materials intended for use as ingredients in such products, such as flavors or fragrances), when subject to the labeling requirements of that Act and the implementing regulations issued by the Food and Drug Administration
3. distilled spirits, wines, or malt beverages intended for nonindustrial use, as defined by the Federal Alcohol Administration Act and implementing regulations, when subject to the labeling requirements of that Act and the regulations issued by the Bureau of Alcohol, Tobacco, and Firearms
4. any consumer product as defined in the Consumer Product Safety Act, or hazardous substance as defined in the Federal Hazardous Substances Act, when subject to the labeling requirement of that Act and the regulations issued by the Consumer Product Safety Commission

Exemptions

Some notable exemptions to the HCS have been established. First, the HCS exempts from compliance some hazardous chemicals present in some mixtures. If a mixture has been tested to ascertain its hazards, the test results provide a basis for determining whether the mixture is hazardous and therefore subject to regulation under the HCS. An untested mixture, however, is regulated according to its components. OSHA presumes that the untested mixture presents the same health hazards as the components that comprise at least 1% (0.1% for carcinogens) of the mixture. Components that comprise less than 1% of the mixture (0.1% for carcinogens) do not trigger the HCS requirements unless there is clear evidence that such an amount could be released in excess of certain regulatory levels or that such a release would present a health hazard to employees.

Second, articles are exempt from compliance with the standard. An "article" is defined as:

> a manufactured item (i) which is formed to a specific shape or design during manufacture; (ii) which has end use function(s) dependent in whole or in part upon its shape or design during end use; and (iii) which does not release, or otherwise result in exposure to, a hazardous chemical, under normal conditions of use.

The stated purpose of this exemption is to exclude from the hazard-communication program items that contain hazardous chemicals in "such a manner that employees won't be exposed to them. . . . Substances inextricably bound in a manufactured item do not present a potential for exposure."

Manufacturers of items that may be considered articles must evaluate their products and the components that are used in their manufacture to determine if any residual amount of a hazardous chemical remains in the article and whether employees of downstream users may be exposed to the chemical. An assessment of the degree of exposure must consider whether the amount released could result in health effects in exposed workers. If so, the effects must be disclosed on the MSDS for the product and the item itself might require a label. OSHA has long used the example of formaldehyde-treated fabric as an example of an item that falls outside the exemption for articles.

Whether or not subsection (iii) of the article definition includes a *de minimis* exemption is unclear. Recently, OSHA considered modifying this subsection to read as follows:

> (iii) which under normal conditions of use does not release more than very small quantities, e.g., minute or trace amounts, of a hazardous chemical (as

determined under paragraph (d) of this section) and does not pose a physical hazard or a health risk to employees.

OSHA stated that the proposed modification "clarifies the definition in accordance with OSHA's enforcement policy to indicate that releases of very small quantities of a hazardous chemical from manufactured items, that do not present a health risk or physical hazard to exposed employees, are not covered by the rule."

Furthermore, in the preamble to a revision of the HCS, OSHA stated:

Releases of very small quantities of chemicals are not considered to be covered by the rule. So if a few molecules or a trace amount are released, the item is still an article and therefore exempted.

Similarly, in its manual on inspection procedures of the HCS, OSHA interpreted the article definition as permitting "the release of very small quantities of a hazardous chemical." If the amount released would not produce health effects in exposed employees, the exemption applies. In the compliance instruction, OSHA states that items may "release . . . very small quantities of a hazardous chemical and still qualify as an article provided that a *physical or health risk* is not posed to employees." Some in OSHA feel that the article exemption is unnecessary and redundant and should be deleted. Many in industry hope that OSHA is moving toward a concept of a *"de minimis* exposure," which implies that exposure assessment is required.

As a practical matter, the definition of the term "very small" is always the determining factor. Each manufacturer must review the scientific evidence to determine at what levels of exposure effects occur. Based on OSHA's interpretation in the recently issued compliance instruction, the definition of very-small quantities turns on whether there is a health or physical risk to exposed employees. If the manufacturer determines that his product is exempt as an article based on the criteria of release of hazardous chemicals, the reasons for the decision should be documented. While there is no requirement that records of such determinations be kept under the standard, the need to prove that the determination was made and the basis for it would suggest that such records be retained. What clearly must be reflected in the manufacturer's written program is a description of the method of determination used.

On the other hand, in *General Carbon Co.* v *OSHA,* OSHA argued, and the court agreed, that an electrical brush that released small quantities such that users were exposed to levels of less than 1.6% of the Permissible Exposure Limit (PEL) of copper dust and less than 6.8% of the PEL of graphite dust was not within the Secretary's interpretation of the article definition because of these hazardous chemical releases; thus the brush was not exempt from the HCS requirements. The court found the Secretary's interpretation of the HCS reasonable.

The Written Hazard-Communication Program

The Standard requires each employer to establish a comprehensive written Hazard-Communication Program for its employees which includes the following: (1) procedures for meeting the Standard's criteria on labels, MSDS, and training; (2) a list of the hazardous chemicals in each work area; (3) procedures for informing employees of the hazards of nonroutine tasks; (4) a list of the hazards associated with the chemicals contained in the pipes in each work area (which should be labeled or coded); and (5) procedures for informing contractors of the hazards to which their employees may be exposed while performing work in the employer's facilities.

Hazard Determination

Paragraph (d), the key to the effectiveness of the HCS, establishes requirements for manufacturers to make a hazard determination for all substances to which employees may be exposed. Paragraph (d)(2) requires manufacturers or employers evaluating chemicals to identify and consider the "available scientific evidence" concerning their hazards. For health hazards, evidence that is statistically significant and that is based on at least one positive study conducted in accordance with established scientific principles is considered sufficient by the Agency to establish a hazardous effect if the results of the study meet the definition of "health hazards" in this section. These studies must be reported on the product's material-safety data sheet. The chemical manufacturer, importer, or employer may also report the results of other scientifically valid studies which tend to refute the findings of a hazard, so long as the MSDS is a fair and accurate representation of the whole of the data known. "The HCS requires that accurate information be provided on the MSDSs. This applies as much to 'over-warning' on the MSDSs/label as well as the absence of information ('under-warning')."

A "health hazard" is defined by Section 1910.1200(c) to mean a chemical for which there is evidence, as defined above, that acute or chronic health effects may occur in exposed employees. The term "health hazard" includes:

Chemicals that are carcinogens, toxic or highly toxic agents, reproductive toxins, irritants, corrosives, sensitizers, hepatotoxins, nephrotoxins, neurotoxins, agents that act on the hematopoietic system, and agents that damage the lungs, skin, eyes, or mucous membranes.

In establishing the definition of health hazards, OSHA identified several sources which presumptively establish that the chemicals listed therein are carcinogenic or otherwise hazardous under the rule. The group of specifically listed substances includes: (1) substances that OSHA regulates; (2) those chemicals identified as carcinogens or potential carcinogens by the International Agency for Research on Cancer or the National Toxicology Program; and (3) substances on the Threshold Limit Value (TLV®) list of the American Conference of Government Industrial Hygienists (ACGIH).

OSHA recognized that most substances to which employees are exposed are not pure chemicals but rather are mixtures. Manufacturers may make the hazard determination of a substance which is a mixture either by testing the mixture as a whole or by assuming that the mixture presents the same health hazards as the individual components which exceed the threshold disclosure concentration (TDC) of the composition. The TDC is set at greater than 0.1% for ingredients meeting the definition of a carcinogen and 1% for all other hazardous ingredients.

Many manufacturers stop at the hazard assessment when they determine that the ingredients present are below the TDC. However, this is not the correct procedure. The standard requires that employers who have "evidence" that a component present below the TDC "could be released in concentrations which would exceed an established OSHA permissible exposure limit (PEL) or and ACGIH threshold limit value (TLV®), or could present a health hazard to employees in those concentrations . . ." should assume that the mixture has the hazards of that ingredient. Evidence of exposure in the use of a product might include data showing that residual amounts are present at some level combined with calculations that a toxicologically significant amount of the ingredient could be present in a workplace under worst-case conditions of use. While exposure calculations are not to be used to determine whether information about ingredients present above the TDC should be included on MSDS and labels, ingredients present below the TDC are not subject to the same evaluation criteria under the standard in the context of making a hazard determination on the mixture as a whole. Thus, it would seem perfectly appropriate to use an exposure-based assessment to determine if the ingredients "present a health hazard to employees in those concentrations. . . ." These provisions are not discussed in the Compliance Instruction issued by OSHA to provide guidance in enforcing the standard, and to some extent manufacturers are required to rely on their own judgment in making the decision on disclosure in these circumstances. The clear thrust of OSHA's oft-repeated statements on disclosure are to include any doubtful information.

If the manufacturer of the mixture obtains information regarding the hazards of a material from a source other than the supplier of the individual substances, the manufacturer is still responsible for including such hazard information on the MSDS and label as appropriate. For an employer who becomes aware of information that is not on the label or MSDS, he must include it in the employee training program required by paragraph (h).

Preparing Labels and Material-Safety Data Sheets

Section 1910.1200(f) of the HCS requires that each container of hazardous chemical(s) in the workplace be labeled with the (a) identity of the hazardous chemical(s) and (b) appropriate hazard warning(s); for chemicals leaving the workplace, (c) the name and address of the chemical manufacturer, importer, or other responsible party must also be included. Of particular importance is the second requirement that the label provide appropriate hazard warnings. OSHA interprets this language to mean that the hazard warning must identify the specific hazard posed by the chemical, meaning that the warning must include statements of target organ effect when they are known. OSHA does recognize that it may not necessarily be *appropriate* to warn on the label about every hazard listed on the MSDS.

Labels and Other Forms of Warning

Chemical manufacturers, importers, and distributors are required to ensure that containers of hazardous chemicals leaving their workplace are labeled, tagged, or marked with the identity of the substance, appropriate hazard warnings, and the name and address of the manufacturer or other responsible party. The labels may not be in conflict with those used to comply with the Department of Transportation (DOT) regulations under the Hazardous Materials Transportation Act.

The Standard requires that each *workplace container* of hazardous chemicals be labeled, tagged, or marked with the identity of the substance inside, and that hazard warnings appropriate for employee protection be included on the label. The hazard warning can be any type of message, wording, picture, or symbol that conveys the hazards of the chemical(s) in the container. The employer is not required to label portable containers into which hazardous chemicals are transferred from labeled containers, and which are intended only for the immediate use of the employee performing the transfer.

As can be seen from the above, OSHA sometimes is very explicit about what is expected. In the HCS, the definition of the written program elements is detailed in very simple language.

> Employers shall develop, implement, and maintain a written hazard-communication program for their workplaces which at least describes how the criteria specified in paragraphs (f), (g), and (h) of this section . . . will be met. . . .

To meet the standard of describing how these requirements are met requires and extensive written program. Simply paraphrasing the language of the standard is not sufficient. OSHA expects to see details such as responsible managers, descriptions, and procedures written out. In its Compliance Instruction, OSHA defined what it expects to see in a written Hazard Communication Program (HCP). It must contain the following elements:

1. The CSHO shall determine whether or not the employer has addressed the issues in sufficient detail to ensure that a comprehensive approach to hazard communication has been developed.
2. In general, the written program should include the following elements where applicable:
   a. *Labels and Other Forms of Warning*
      i. designation of person(s) responsible for ensuring labeling of in-plant containers
      ii. designation of person(s) responsible for ensuring labeling on shipped containers
      iii. description of labeling system(s) used
      iv. description of written alternatives to labeling of in-plant containers, where applicable
      v. procedures to review and update label information when necessary
   b. *Material-Safety Data Sheets*
      i. designation of person(s) responsible for obtaining/maintaining the MSDS
      ii. how such sheets are to be maintained (e.g., in notebooks in the work area(s), via a computer terminal, in a pickup truck at the job site, via facsimile) and how employees obtain access to them
      iii. procedure to follow when the MSDS is not received at the time of the first shipment
      iv. for chemical manufacturers or importers, procedures for updating the MSDS when new and significant health information is found
   c. *Training*
      i. designation of person(s) responsible for conducting training
      ii. format of the program to be used (audiovisuals, classroom instruction, etc.)
      iii. elements of the training program—compare to the elements required by the HCS (paragraph (h))
      iv. procedures to train new employees at the time of their initial assignment and to train employees when a new hazard is introduced into the workplace
      v. procedures to train employees of new hazards they may be exposed to when working on or near another employer's work site (i.e., hazards introduced by other employees)
      vi. guidelines on training programs prepared by the Office of

Training and Education entitled "Voluntary Training Guide-lines" (49 *Fed. Reg.* 30290, July 27, 1984) can be used to pro-vide general information on what constitutes a good training pro-gram

d. *Additional Topics to Be Reviewed*

   i. Does a list of the hazardous chemicals exist and if so, is it com-piled for each work area or for the entire work site and kept in a central location?

   ii. Are methods the employer will use to inform employees of the hazards of *nonroutine* tasks outlined?

   iii. Are employees informed of the hazards associated with chemi-cals contained in unlabeled pipes in their work areas?

   iv. Does the plan include the methods the employer will use at multiemployer work sites to inform other employers of any pre-cautionary measures that need to be taken to protect their em-ployees?

   v. For multiemployer workplaces, are the methods the employer will use to inform the other employer(s) of the labeling system used described?

   vi. Is the written program made available to employees and their designated representatives?

The HCS therefore imposes a comprehensive duty on employers to develop written documents both explicitly and implicitly.

Material-Safety Data Sheets

The Standard requires chemical manufacturers and importers to develop material-safety data sheets (MSDS) for each hazardous chemical they produce or import. Employers that use such substances are required to obtain MSDS for each haz-ardous chemical used in their workplaces from their suppliers. Alternatively, an employer may develop its own MSDS for substances obtained from suppliers. Employers that prepare hazardous chemical mixtures or compounds for use within the plant must develop MSDS for such substances. The Standard specifies the information that must be provided on the MSDS.

All MSDS must be in English, and include the identity, as well as the chem-ical and common names, for the hazardous chemical. The MSDS must also pro-vide the information specified in the Standard on the physical and chemical char-acteristics of the hazardous chemical, the known acute and chronic health effects and related health information, information concerning exposure limits, whether the chemical is considered to be a carcinogen by recognized authorities, precau-tionary measures, emergency and first-aid procedures, and an identification of the person responsible for the sheet. Special provisions apply when listing ingredients for hazardous chemical mixtures.

Each MSDS received from our chemical suppliers must contain the following information:

1. the chemical identity used on the label
2. for single substance—the chemical and common names of the substance
3. for mixtures—the chemical and common names of the mixture and the ingredients that contribute to the hazardous nature of the mixture
4. the physical and chemical characteristics of the hazardous chemical
5. the physical hazards of the hazardous chemical, including the potential for fire, explosion, and reactivity
6. the health hazards of the hazardous chemical, including signs and symptoms of exposure, and any medical conditions that are generally recognized as being aggravated by exposure to the chemical
7. the primary route(s) of entry
8. the OSHA permissible exposure limit, ACGIH threshold limit value, and any other exposure limit used or recommended
9. whether the hazardous chemical is listed in the National Toxicology Program (NTP) "Annual Report on Carcinogens" latest edition, or has been identified as a potential carcinogen in the International Agency for Research on Cancer (IARC) Monographs, latest edition, or by OSHA
10. any generally applicable precautions for safe handling and use, including appropriate hygienic practices, protective measures during repair and maintenance of contaminated equipment, and procedures for cleanup of spills and leaks
11. any generally applicable control measures such as appropriate engineering controls, work practices, or personal protective equipment
12. emergency and first-aid procedures
13. the date of (i) preparation of the material-safety data sheet or (ii) the latest revision to it
14. the name, address, and telephone number of the chemical manufacturer, importer, employer, or other responsible party preparing or distributing the MSDS, who can provide additional information on the hazardous chemical and appropriate emergency procedures, if necessary

MSDS may be developed in either of two formats. The first alternative is a single MSDS. The second alternative is to attach the component chemical MSDS received from the suppliers to a cover sheet identifying the mixture, listing all hazardous components and referencing the attached MSDS for further information. Again, the second alternative is acceptable only if the hazards associated with the mixture are not different from those of its components and the associate preparing the MSDS is not aware of any additional information which should be included.

The firm that prepares the MSDS must ensure that it accurately reflects the scientific evidence that forms the basis for the hazardous chemical determination. Any significant new information that becomes available to the preparing firm concerning the potential health hazards of the chemical or ways to protect against the hazards must be added to the MSDS within three months of discovering the information. In-plant labels that refer to these MSDS should also be updated accordingly.

Chemical manufacturers must provide an MSDS to purchasers of hazardous chemicals in advance of or with their first shipment. Manufacturers that use hazardous chemicals must maintain copies of the MSDS, and ensure that they are readily accessible to their employees during each work shift. MSDSs must be made available to employees and their designated representatives.

For example, in its Compliance Instruction, OSHA states, in pertinent part, as follows:

> [T]he selection of hazards to be highlighted on the label will involve some assessment of the weight of the evidence regarding each hazard reported on the data sheet. Assessing the weight of the evidence prior to including a hazard on a label will also necessarily mean consideration of exposures to the chemical that will occur to workers under normal conditions of use, or in foreseeable emergencies.

OSHA further amplifies on this guidance by citing ethanol as a chemical with known carcinogenic effects *via* ingestion, but not by the inhalation or dermal routes of exposure. Thus, OSHA concludes that while the MSDS for ethanol must disclose its carcinogenic potential, the label need not include a cancer hazard warning where the product's intended use does not involve exposure through ingestion.

Section 1910.1200(g) requires that manufacturers are to obtain or develop a material-safety data sheet for each hazardous chemical they produce or import. MSDS are required to state if the hazardous chemical is a mixture that has been tested as a whole to determine its hazards, and the chemical and common names of the ingredients that have been determined to be health hazards, and that exceed the TDC.

The HCS also requires manufacturers to update MSDS and labels as new information on existing chemicals becomes available. The person who prepares a MSDS revise the data sheets within three months of becoming "newly aware of any significant information regarding the hazards of a chemical . . . " must prepare a revised MSDS and provide distributors and customers with them with the first shipment after the MSDS is updated, and must include it with the shipped containers unless it is provided to the downstream employer prior to or at the time of shipment. As a matter of prudence, the MSDS should be transmitted to distributors with a reminder to distribute them to their downstream customers in accordance with these requirements.

Employee Information and Training

Employers must establish a training and information program for employees exposed to hazardous chemicals in their workplace. The training must be provided at the time of initial assignment, and whenever a new hazard is introduced to the work area. This includes newly developed or discovered physical or health-hazard data on existing chemicals in the work area. Employees must be informed of the Standard's requirements, and the components of the employer's compliance program. Employees are to be informed of operations in their work area where hazardous chemicals are present, and the location where the employer maintains the written materials required under the Standard (i.e., this document and MSDS for workplace hazardous substances).

Employees must be trained in safe work practices, how to detect the presence of hazardous chemicals in their work area, information about physical and health hazards of chemicals in the work area, how to perform nonroutine tasks safely, how to determine hazards by reading a label, how to select and use personal-safety equipment, how and where to obtain copies of the MSDS and ensure that they are accessible to their employees during each work shift. MSDSs must be made available to employees and their designated representatives.

Trade Secrets

The OSHA Standard defines trade secrets as "any confidential formula, pattern, process, device, information, compilation of information that is used in an employer's business, and that gives the employer an opportunity to obtain an advantage over competitors who do not know or use it." It provides that a chemical manufacturer, importer, or employer may withhold information, such as the specific chemical identity of a hazardous chemical, from a material-safety data sheet if: (1) the claim that the information withheld is a trade secret can be supported, (2) the MSDS contains information on the properties and effects of the hazardous chemical, (3) the MSDS indicates that the specific chemical identity is being withheld as a trade secret, and (4) the specific chemical identity is made available to health professionals, employees, and designated representatives in accordance with the special provisions of the Standard dealing with trade secrets which are outlined below. Though universities and hospitals are not traditional owners of trade secrets, future patentability of an innovative development may depend on the ability of the chemist or physician to protect his or her chemical development against pre-patent disclosure.

Where a treating physician or nurse determines that a medical emergency exists and the specific chemical identity of the hazardous chemical is necessary for emergency or first-aid treatment, the information must be released regardless of whether the disclosure is covered by a confidentiality agreement.

In nonemergency, health-related situations, the employer is required to disclose, to a health professional, employee, or designated representative, a specific chemical identity otherwise permitted to be withheld as a trade secret if a request for that information: (1) is in writing, (2) describes in reasonable detail the occupational health need for the information, (3) explains why the disclosure of the specific chemical identity is essential, and why disclosure of alternative information is inadequate, (4) includes a description of the procedures to be used to maintain the confidentiality of the information, and (5) the health professional, employee, or designated representative agrees in a written confidentiality agreement that the trade-secret information will not be used for any purpose other than the health need.

The chemical manufacturer, importer, or employer may deny a written request for disclosure of a specific chemical identity. OSHA then has a right to review the denial. If OSHA determines that the specific chemical identity requested is *not* a bona fide trade secret or is a trade secret but the party requesting the information has a legitimate medical or occupational health need for it, then OSHA may direct disclosure and impose limitations or conditions upon the release of such information so that the health services are provided while minimizing the risk to the employer. Failure to comply with such a directive may result in a citation.

Each MSDS should reflect the scientific evidence examined by the chemical manufacturers in making the hazard determination. The chemical manufacturers are required to update their MSDS as new information becomes available and to forward the updated MSDS for a particular chemical with the first shipment of the chemical following the revision of the MSDS.

If the preparing firm becomes aware of any significant information regarding the hazards of a chemical or ways to protect against the hazards, this new information must be added to the MSDS within three months. The company must be supplied with the updated MSDS with the first shipment after the MSDS is updated.

Labels should also be updated to be consistent with new MSDS. Whenever a new hazard is introduced into the work area, workers should be apprised of newly discovered physical or health hazards or ways to protect against the hazards of chemicals in the work area. To the extent newly discovered hazards appear in revised MSDS, workers should be informed and trained accordingly.

If the substances in the container are regulated by OSHA in a substance-specific health standards, the label must contain the specific language required by that standard. Employees must not remove or deface existing labels on incoming containers of hazardous chemicals unless the container is immediately relabeled with the necessary information. If bulk packages are broken down within the facility, the employer must assure that the individual packages or containers are relabeled. There is a limited exemption from the labeling requirement for portable containers into which hazardous chemicals are transferred from labeled containers if the chemical will be completely used by the person who performs the transfer from a labeled container during the same shift in which the transfer is made.

Communication with Contractors

Host employers must take steps to ensure that contractors working in their facilities are advised of the presence of hazardous chemicals to which the contractor employees may be exposed. Likewise, agreements with contractors must require the contractor to advise the host employer of the hazardous chemicals which the contractor will bring into the plant. Based on the information provided by the contractor, the host employer must determine whether employees or other contractor's employees may be exposed to the hazardous chemicals, and what additional information concerning the contractor's activities must be communicated to employees and to other contractors. The host employer is responsible for assuring that the contractor has sufficient information to inform the contractor employees and protect them from exposure to hazardous chemicals used in the facility or brought into the plant by other contractors. The information provided must include: (1) where the MSDS for each chemical presenting an exposure potential for any and all employees may be found, (2) any precautionary measures that need to be taken during normal operating conditions and in foreseeable emergencies, and (3) a description of the labeling system used.

9.10   COVERAGE OF LABORATORIES

Laboratories are subject to limited coverage under the HCS. Much of their regulation comes from 29 C.F.R. §1910.1450, the laboratory standard. Container labels on incoming shipments of hazardous chemicals must not be removed or defaced unless immediately replaced, and MSDS which are received for such substances must be maintained and made available to laboratory personnel. Laboratory personnel must follow good laboratory practices and should label all containers of hazardous chemicals, other than portable containers whose contents will be used immediately, with the identity of their contents.

Laboratory personnel must receive training and instruction concerning the HCS and the hazardous chemicals with which they may have contact and (1) safe methods of handling chemicals in laboratories, (2) physical and health hazards of the chemicals generally, and (3) protective measures that may be relevant to such chemicals, including specialized protective equipment and work practices.

9.11 CITATIONS

General Duty Clause and Citations

Under the statute, OSHA may issue a citation to an employer for failure to comply with a particular health or safety standard, or in the case of serious health or safety hazards for which there is no specific standard, a violation of the General Duty Clause (GDC). The General Duty Clause of the statute, Section 5(a), requires the employer to

> furnish to each of his employees employment and a place of employment which are free from recognized hazards that are causing or likely to cause death or serious physical harm to his employees.

Thus, by statute, before OSHA can cite an employer under the GDC, there must be a danger threatening physical harm to employees who are exposed to the danger. The lack of a particular abatement method or the occurrence of an accident are *not* violations of the GDC; they may be evidence of such a violation, but additional elements must be proven.

First, the hazard must be recognized, either by the employer specifically, by the industry generally of which the employer is a part, or by "common sense." In its Field Operations Manual, OSHA identifies nine different ways to establish that a hazard is recognized by an industry. Among them are statements by health and safety experts familiar with the industry; implementation of an abatement method; warning labels on equipment; studies conducted by employers, unions, or trade associations; state or local laws; and consensus standards from organizations such as the National Fire Protection Association (NFPA) and the American National Standards Institute (ANSI).

Similar sources may be used to demonstrate that the individual employer recognizes the hazard. Company memoranda, safety rules, inspections, operating manuals, safety audits, or accident reports are a few examples of the kinds of documents that can be used. Employee complaints and halfhearted or ill-maintained corrective actions also may show that the employer was aware of the hazards. Finally, where any reasonable person would recognize the hazard, OSHA may cite under the GDC.

The second element of the GDC violation is that the hazard must be likely to cause death or serious physical harm, defined by OSHA as

Impairment of the body in which part of the body is made functionally useless or is substantially reduced in efficiency, and which may be temporary
or permanent, chronic, or acute, usually requiring treatment by a physician.

Cuts, lacerations, or punctures involving significant bleeding and/or requiring sutures are considered by OSHA to be serious physical harm, but abrasions,
bruises, or contusions generally would not be.

Finally, OSHA must be able to show that the hazard may be abated by a feasible and useful method. The focus is on whether the recognized hazard is preventable, consistent with the Congressional purpose. Legal precedent requires
that the GDC may be used only where a specific standard is not available, unless
the employer knows the particular standard is inadequate to protect his employers
from the hazard.

Repetitive-Motion Injury

Most recently, OSHA has used the GDC to cite employers where a substantial
number of employees suffer from repetitive-motion injury, sometimes called
Cumulative Trauma Disorders (CTD). In the meat-packing and poultry industries, in automotive and food processing, and in a number of other cases, OSHA
has shown that employers failed to take necessary steps to prevent the occurrence
of conditions such as carpal tunnel syndrome. In settlements involving millions
of dollars in fines, OSHA has imposed extensive programs of investigation, education, engineering, and medical surveillance. Some believe that OSHA has followed this path because of frustration with the standards-setting process, or because the Agency is unable to articulate a scientifically defensible and ultimately
enforceable standard. Arguably, establishment of a standard of care in this manner, under which employers are liable for citation and imposition of the full force
of OSHA's enforcement authority, fails to provide adequate notice to the regulated community and opportunity for comment as required by the standards-setting authority of the statute. This process of stealth rule making ultimately un-
dermines OSHA's persuasive authority.

Statute of Limitations and Timing of Citations

Citations must be issued within six months of date of inspection. *The six-month
period begins to run on first date of inspection, not when OSHA investigation is
complete.* When OSHA amends the original citation, the amendment relates back
to the original citation date and is not barred by the fact that more than six months
have passed between inspection and amendment.

Certain citations may be issued immediately at discretion of the inspector, although the CSHO has no authority to issue proposed penalties at the work site.

Factors which the CSHO must consider include: whether laboratory analysis is required; whether the advice of specialists or legal counsel is needed; and whether OSHA's jurisdiction is in question. The CSHO must discuss his or his findings with the OSHA area director before issuing work-site citations.

Citations issued after the closing conference must be issued "with reasonable promptness after termination of the inspection." The Occupational Safety and Health Review Commission (OSHRC) has rarely vacated a citation because of delay in issuance as long as it is issued within six-month statute of limitations, unless actual prejudice to the employer's preparation of a defense against the citation could be shown. A lack of "reasonable promptness" is an affirmative defense and must be pleaded and proved by employer. Citations are mailed by certified mail, return receipt requested. The date of the return receipt defines the beginning of the contest period. A notice of proposed penalty must be given to someone who has authority to pay the violation or contest the citation. Service of process must be made on corporate official, not employee at the work site. The Review Commission will normally find adequate service when it is reasonably calculated to give the employer actual notice, even if not given to the proper person.

Content of Citations

The citation must "describe with particularity the nature of the alleged violation, including a reference to the provision(s) of the Act, standard, rule, regulation, or order alleged to have been violated." The employer must be able to identify the violation although citation does not have to state all details of the alleged violation. Lack of particularity is an affirmative defense to be pleaded and proved by employer. The citation must include proposed penalty and must fix a reasonable time for abatement; it may be amended before or after filing of notice of contest if additional facts establish a need for revision.

The employer must post the citation or a copy of it at or near each place an alleged violation occurred or in a prominent place for all affected employees to see. The citation must remain posted until the violation is abated or for three working days, whichever is later.

9.12   OSHA Penalties

Citations are classified by OSHA according to their seriousness. *De minimis* violations are those involving a technical violation of the standard which does not result in a hazard to health or safety. An example of such a violation might be the installation of a stair rail at a height slightly above the maximum permitted under the standard, or the use of a nuisance-dust respirator against a nontoxic dust without a respiratory-protection program. Such violations are noted, but carry no precedential value, including the fact that they cannot be used as the basis for a repeat violation. No abatement period is fixed for correction of the condition, and no penalties are assessed.

Other-than-serious violations are those for which the risk of serious physical harm is small or the likelihood of injury is so remote as to be nearly impossible of occurrence. *Serious* violations are those that involve the potential for real harm to employees. The penalty for serious violations can be as much as $7,000, while willful or repeated violations carry potential fines of up to $70,000 per violation.

These categories are not fixed, and similar conditions can be classified as more serious violations as OSHA gains more experience, as is illustrated by citations issued for violations of the Hazard Communication Standard (HCS). Between 1988 and 1990, citations issued for serious violations of the HCS increased by a factor of six, from 1,700 to over 10,000. Other-than-serious citations dropped from nearly 14,000 to approximately 6,000. While OSHA surely would argue that the standard has not changed, it is clear that the Agency's interpretation of the standard has changed, at least with respect to severity. This is amply demonstrated by the fact that over 60% of the violations in 1990 were categorized as serious, as compared to less than 10% two years earlier.

9.13   APPEALS OF CITATIONS

Employers are permitted initially an appeal under the OSH statute, pursuing an administrative informal process. An employer, or employee representatives, may request an informal conference upon receipt of citation and notice of proposed penalty. Requests for these conferences are granted as matter of routine before the end of the 15-working-day contest period. The conference is held with the OSHA area director or his representative with authority to negotiate a agreement to settle the citation without litigation. Area directors are authorized to enter into informal settlement agreements only before the employer files a written notice of contest (NOC) (see below). The informal conference provides an opportunity (1) for resolving disputed citations and penalties without resorting to litigation, (2) to improve understanding of specific safety or health standards that apply, (3) to discuss ways to correct violations, and (4) to discuss problems with proposed abatement dates.

The benefits of informal conferences include rapid settlement of a citation and the opportunity for the employer to obtain a reduction in the penalty or modification of the abatement date. Neither the informal conference nor the request for a conference stays the 15-working-day period within which employer must file the NOC or for filing notice alleging that the time fixed for abatement is unreasonable. Employers should consider filing the NOC as contingency even if they choose to attempt to negotiate a settlement.

Notices of Contest

A notice of contest (NOC) is the beginning of the formal appeals process. The procedure is governed by regulation and is conducted before the Occupational Safety and Health Review Commission (OSHRC). The benefits of contesting citation include: possible dismissal or modification of the violation; possible reduction of proposed fines; and possible modification of abatement. The employer should be aware that acceptance of certain citations (for example, willful citations) may support criminal charges. The main disadvantage of contesting citation is the substantial cost involved in litigation.

While the Act itself does not specify the procedures for initiating the NOC, the implementing regulations state that the employer must notify the area director "in writing." The Review Commission has found oral Notice of Contest to be valid in some cases; however, the general rule is that the NOC must be in writing. No particular form is required for the NOC and the Review Commission is liberal in its interpretation of documents that satisfy the requirement for giving notice. However, the thoughtful employer will prepare the NOC carefully so that it makes clear which items of the citation are being contested (for example, if OSHA finds that only the amount of the penalty is contested, the employer will not be allowed later to contest the citation itself). A copy of both the citation and the NOC must be posted at the work site where it is accessible to affected employees, allowing employee representatives to file an NOC to challenge the reasonableness of an abatement date. A Notice of Contest may be withdrawn at any stage of the proceedings.

The NOC must be postmarked within 15 working days of receipt of citation. Employers are nearly always unsuccessful in seeking extension of 15-day period. An employer may be excused for not filing within the 15-day period if the delay was caused by reliance on erroneous representations, deception, or failure by OSHA to follow proper procedure; however, except in the most unusual cases, his own delays will not relieve the employer of filing the NOC within the statutory period. A failure to file timely NOC will convert the citation into a final order not subject to review and will subject employer to follow-up inspection to determine whether the violation has been corrected.

Extending the Abatement Period

The burden of showing reasonableness in the abatement period is on OSHA. A Petition for Modification of Abatement (PMA) may be filed when an employer has made a good-faith effort to comply, but abatement has not been completed due to factors beyond his control, or when abatement would cause a significant financial hardship to the employer. The PMA must be filed no later than the next working day following the date on which abatement originally required. It must be in writing and include the following: (1) steps already taken in an effort to comply and the dates those steps were taken, (2) the amount of additional time needed to comply, (3) the reasons the additional time is needed, (4) interim steps being taken to protect employees, and (5) certification that the petition has been posted in a conspicuous place in the work site. If uncontrollable circumstances prevent the employer from meeting the abatement date and the 15-working-day contest period has expired, a PMA may be filed if accompanied by a statement explaining the circumstances.

Once an NOC is filed, the abatement period does not begin to run until the entry of the final order by the Review Commission. If the Review Commission finds that the only purpose of the employer's Notice of Contest is to delay or avoid the abatement period, the abatement period begins to run from date of receipt of the citation.

Opportunity for Settlement After Filing Notice of Contest

Settlement is encouraged at all stages of proceeding as long as it is consistent with provisions and objectives of the act. The employer may not withdraw his Notice of Contest in favor of settlement unless there is certification of date on which violations have been or will be abated. Settlement is treated as an admission of liability in the original citation unless a settlement clause provides otherwise. In addition, the settlement agreement may be used in future proceedings concerning subsequent violations to show knowledge of a condition or practice.

Under the rules governing settlements, an administrative law judge (ALJ) is assigned when "there is a reasonable prospect of substantial settlement with assistance of mediation by a settlement judge." Affected employees who have party status must receive written notice of a proposed settlement (either by personal service or posting) so they can file objections with the ALJ within ten days.

Appeals Process

From the OSHRC, employers or OSHA may appeal to a circuit court of appeals. The process beyond the Review Commission is identical to any other appeal of a judicial proceeding. Substantively, however, the standard of review is a more lenient one. Unless unreasonable or inconsistent with the standard or the statute, the interpretation of the Secretary of Labor is controlling. There must be substantial evidence in the record taken as a whole to support the Secretary's interpretation, and the factual determinations of the OSHRC are final.

REFERENCES

Association of American Railroads v *DOT*, 38 F.3d 582; 1994 U.S. App. LEXIS 29993 (D.C. Cir. October 28, 1994).

Martin, Secretary of Labor v *Occupational Safety and Health Review Commission (OSHRC)*, 59 U.S.L.W. 4197, 111 S. Ct. 1171 (1991).

Chapter 10

Pollution Prevention:
Minimizing Costs and Liability

10.1 INTRODUCTION

The improper management of solid waste streams can result in serious financial burdens and legal risks. Many opportunities exist for institutions to reduce or, preferably, prevent pollution at the source through cost-effective modifications in operations, use of alternate materials, and improved waste-stream management. Solid-waste recycling from the institution's routine garbage stream is the simplest and easiest way to start (O'Reilly 1992).

Until the mid-1980s, institutions were not overly concerned with the costs of pollution resulting from solid-waste generation. The Resource Conservation and Recovery Act (RCRA), amended in 1984 by the Hazardous and Solid Waste Amendments (HSWA) was specifically written with heavy industry in mind, placing a heavy burden on insititutions that were not as equipped as industry was to deal with paperwork, fines, and control systems (O'Reilly, Kane, and Norman 1993). These RCRA hazardous-waste regulations make it difficult and costly for an institution to properly dispose of hazardous wastes. Chemical hazardous wastes are generated through teaching, research, patient-care services, and day-to-day operational activities conducted at various institutions. That set of wastes is the primary area of attention for educational and health-care institutions' waste-reduction and pollution-prevention programs.

10.2 REDUCTION OF HAZARDOUS WASTES

Normal solid waste streams should be minimized using whatever techniques are feasible, and hazardous waste even more so should be managed with reduction and minimization as a goal. Inside the institution, this desire to reduce waste may collide with the experimental spirit of discovery that chemists, physicians, and other scientists desire, but it is the best means to success for the institution in a climate of regulatory costs and penalties.

Definition of a Hazardous Waste

As Chapter 8 describes at length, the Environmental Protection Agency (EPA) regulates hazardous wastes under the Resource Conservation and Recovery Act (RCRA) through Title 40 of the Code of Federal Regulations, Parts 260–270. RCRA regulates the identification, generation, transportation, storage, and disposal of hazardous wastes, by persons including nonprofit health-care and educational institutions. RCRA also establishes a severe penalty structure (up to $50,000 per day for each offense, determined by a penalty matrix approach) and allows for criminal prosecution (jail sentences of up to five years) of those individuals and organizations who improperly manage hazardous waste. Transportation of hazardous waste, as discussed in Chapter 7, is regulated by the Department of Transportation (DOT) in Title 49 of the Code of Federal Regulations, Parts 171–173.

"Hazardous" waste is defined as any waste or combination of waste which because of its quantity, quality, concentration, physical, chemical, or infectious characteristics could cause or significantly contribute to adverse effects in the health and safety of humans or the environment if improperly managed. This includes any wastes exhibiting a *general characteristic of ignitability, corrosivity, reactivity,* or *toxicity* (according to toxicity characteristic leaching procedure (TCLP) testing). Also included are all wastes *specifically listed* by the EPA in 40 CFR Part 260 to 270 as a *toxic* or *acutely toxic* hazardous waste, or wastes derived from *specific* or *nonspecific sources.*

General Characteristics

The following guidelines should be followed when determining if a chemical waste exhibits one or more of the general characteristics:

1. **Ignitability** (EPA Code D001)
   a. liquids with a flash point of less than 140°F (60°C)
   b. solids capable of causing fire spontaneously, by friction, or absorption of moisture, and when ignited, burn vigorously and persistently to create a fire hazard
   c. flammable compressed gases
   d. oxidizers (yield oxygen readily to stimulate combustion)

2. **Corrositivity** (EPA Code D002)
   a. aqueous solutions with pH equal to or less than 2 or greater than 12.5
   b. liquids capable of corroding steel at a specified rate and temperature
3. **Reactivity** (EPA Code D003)
   a. substances that react violently with water or air, or produce toxic gases or explosive mixtures when mixed with water or air
   b. substances that are normally unstable or explosive
   c. chemical compounds containing cyanide or sulfide
4. **Toxicity** (EPA Code D series)
   a. materials that fail the TCLP test because of the presence of certain heavy metals or organic constituents above regulated levels

Specifically Listed Wastes

Approximately 500 chemicals are specifically listed by EPA as hazardous waste:

5. **P Listed Waste**s (EPA code P), acute toxics
6. **U Listed Wastes** (EPA code U), toxics
7. **F Listed Wastes** (EPA code F), wastes from nonspecific sources
8. **K Listed Wastes** (EPA code K), acute toxics

   If your institution generates hazardous wastes, then an Environmental Protection Agency (EPA) generator number must be obtained using EPA Form 8700-12. This is basically a notification of hazardous-waste activity. It is sent to the entity having jurisdiction in your particular case (EPA or state program with authority to administer their own hazardous-waste program). If an institution does not have in-house capability to determine if a generator number is needed, then any of the many hazardous-waste-disposal contractors would be glad to help you make a determination.

10.3   SOLID-WASTE REDUCTION[1]

Recycling provides the most efficient use of our finite natural resources. It means the creation of new industry, employment opportunities, research, advanced technology, and a reduction of imported natural resources. At this moment Japan,

---

[1] Adapted with permission from the Georgetown University recycling information booklet prepared by Jon Miller, Copyright Georgetown University 1995.

Scandinavia, and Western Europe are researching and investing in more efficient techniques to reduce the economic and environmental costs of production. Reducing pollution sources and their adverse impacts will require a balance between resource allocation, production, and consumption.

Increased costs of disposing of institutional solid-waste streams can be be minimized through the use of an effective recycling solid-waste program. Use of an effective program to manage the solid and hazardous-waste streams can result in reduced operational costs and good public relations with the local community.

There are many ways to reduce waste by recycling solid-waste streams in an institutional setting:

Office Paper

- The energy used to produce one day's junk mail in America is enough to heat 250,000 homes.
- The average American uses 600 pounds of paper per year.
- Each ton of recycled paper saves 17 trees.

Also known as ledger, office paper is one of the major by-products of any institution. Recycling old paper instead of using new timber to produce paper uses 60% less energy and 61% less water, and creates 74% less air pollution and 35% less water pollution. Recycling is vital to the saving of our forests, for even with the massive reforestation operations in the United States and the rest of the world, there is no way we can continue to support the current rate of consumption of virgin timber. If you recycled all the paper that you used in one year, you would personally save five trees. Now multiply by any population count you choose. That's a very good start at pollution minimization.

Recycling is not new to the paper industry; in fact, waste-paper recycling has existed in the United States for over 300 years. When paper began to be milled in the United States in 1690 it was entirely made up of recycled fiber from used textiles. It wasn't until almost 200 years later, when the demand for paper products increased drastically, that wood fiber became an integral part of the papermaking process.

Office paper comes in many different grades, each of which is of different value to processors, with white office paper being of the highest value. Two other commonly used grades are mixed office paper and nonledger mixed paper, with the latter being the lowest grade. It is important to keep each grade of paper separate from the others. For example, *one sheet* of nonwhite paper, *one* plastic-windowed envelope, or *one* Styrofoam cup can downgrade the value of an entire *ton* of white office paper. This directly reduces the institution's income from sale of the recycled, sorted waste materials—so it is worth the effort to do the separation.

White Office Paper

White office paper includes the following: greenbar or white computer printout (no carbon-copy printout), laser-printer paper, white xerographic paper, folder paper, white spiral notebook paper (the cover must be removed), white index cards, white bond paper, and adding machine tape.

Mixed Office Paper

Also called filestock, this includes colored paper, windowless envelopes, colored index cards, unbleached recycled paper, file folders, manila cardstock folders, rag bond, and this book.

Mixed Paper

Mixed paper may be construction paper, kraft envelopes, wrappers of paper reams, onionskin typing paper, Post-It® notes, crêpe paper, posters, calendars, playbills, NCR forms (noncarbon record paper), gift-wrapping paper, folders, Rolodex® cards, and business cards.

When recycling, it is *extremely* important that your white paper be free of anything not listed in the white-office-paper category. The following section is a troubleshooting guide for paper; the information is applicable to the following section on newspaper as well.

Unacceptable Contaminants

- *Cardboard:* Cardboard is constructed of many layers and is much thicker than normal paper, and it has its own recycling process. For more information, see the "cardboard" section below.
- *Paper bags:* Paper grocery bags are made of an extremely low-grade material, and presently there is no processor in the area willing to take this material. Although many paper bags are made from recycled fibers, they are made from generally poor materials or have been in the recycling chain too long to maintain their fiber quality through repeated processing.
- *Plastics:* Whether in the form of spiral or pressed binding on a report or the bag you collected your recyclable paper in, plastics are bad for paper mills. They are hazardous to paper-processing plants because in the pulping process the plastic melts and sticks to machinery parts, requiring costly

shutdowns and repairs. *Plastics are the worst possible thing that can be put into a paper recycling bin!*

- *Glues and tapes:* Plastic tape that is attached to paper products often goes unnoticed, glue coatings even more so. The buildup of tape and glue are just as dangerous as putting a plastic bottle directly into the paper receptacle. Please ensure that any glued or taped paper and glue binding from notepads does not make its way into the recycling stream.
- *Stickers:* A sticker is nothing more than decorative tape that doesn't hold anything together.
- *Waxed paper and paper cups:* Wax has similar effects on plant machinery to those of plastics, and there is no current process to remove the wax from the paper before or after pulping.
- *Carbon paper:* Carbon paper is very similar to waxed paper, but the paper is of a lower grade and is saturated with the carbon ink, so not all of the ink can be removed during the de-inking process. During processing the remaining ink will streak the other pulp in the vat.
- *Food matter:* Food is a *definite* negative in paper collections! The worst kind of food matter is cheese or any other really viscous material, as it exhibits the exact same coagulent qualities as those of plastics and wax to clog the processing system.
- *Tissue paper and paper napkins:* These products are of an extremely low paper grade and low fiber content. In addition, they usually contain contaminants in the form of food matter or other waste. Regrettably, all these contributing factors make them absolutely nonrecyclable.
- *Coffee filters:* Coffee filters are practically identical to tissue papers and paper napkins, and then there is the matter of the coffee grounds that were in them.
- *Federal Express and overnight mail envelopes:* These are either made out of plastic or have a thick plastic coating. *Not recyclable.*
- *Protective envelopes:* Any envelopes that are lined with plastic bubble wrap or filled with reused newspaper packing material are destined for the trash.
- *Envelopes with plastic windows:* These envelopes can boost their status by the simple act of you tearing out the windows, making them filestock bound.
- *Photographs:* Photographic paper is not recyclable, and has a plastic coating.

Newspaper

- The average American uses 120 pounds of newsprint per year.
- It takes an entire forest of 500,000 trees to supply all of America with the Sunday edition of the newspaper.

- If every newspaper printed in the United States alone was recycled, 250 million trees would be saved each year.

Newspaper is the primary paper product collected from households. Newsprint comprises fully 14% of solid waste in landfills, but it is not as degradable as some might think. Garbage archaeologists have excavated newspapers from the 1950s that are just as legible now as the day they were printed. The high volume of newspaper recycling is very efficient in saving landfill space and turning old papers back into new newsprint. Most paper can be recycled six to seven times into different grades of mixed paper goods. Newspaper can also be made into fiberboard, packing material, insulation, roofing materials, animal bedding, and many other useful products.

Many people don't bother to recycle newspaper because they refuse to remove the glossy inserts from the bulk of the newsprint. However, the glossy inserts are recyclable as newspaper and actually are beneficial to some processors' recycling process. Glossy magazines can also be recycled.

Unacceptable Materials in Newspaper Recycling Bins

- *Plastics:* Some people receive their newspaper and magazine subscriptions wrapped in plastic. Some of these people never actually read their newspapers or magazines and just toss them, complete with plastic, into the recycling bin. Remove the plastic packaging and then deposit the newspaper or magazine for recycling. Remove the rubber bands as well.
- *Food matter:* Food matter in the form of solid food or beverage can destroy the value of newspaper for recycling. In fact, newspaper that is wet simply by water is often rejected by recyclers.

Cardboard

- The recycling rate of cardboard is exceeded only by aluminum cans.

The largest quantity of recycled paper grade comes from corrugated-cardboard containers.

Unacceptable Materials in Cardboard Recycling Bins

- *Pizza boxes:* This is often the largest category of cardboard at an institution that runs 24 hours a day. There is always food contamination on them, especially cheese, which renders the cardboard useless for recycling.

- *Adhesives:* Old, reused boxes are likely to have layers and layers of old packing tape on them. Small amounts are acceptable, but if the tape is excessive, either completely remove any tape or stickers before recycling, or throw the boxes away.
- *CD boxes, cereal or powdered consumer product boxes, egg cartons, and milk cartons:* These are not actually cardboard products. They are made out of a material called pressboard which is several pressed or molded layers of a low-grade paper product. Although recyclable, there are very few processors willing to accept it due to the complete absence of a market for recycled pressboard. In addition, milk cartons and the bags inside cereal boxes are plastic or waxed paper and thus unacceptable.
- *Plastics:* Packing tape is not the only source of plastics you might find in your corrugated boxes. Some corrugated boxed are actually made completely out of plastic, but just because they're corrugated doesn't mean that they can be recycled with the rest of your corrugated boxes.
- *Foam packing peanuts:* Any polystyrene foam products are *highly* unacceptable for paper-recycling purposes.

Telephone Directories

Phone books are something that everyone has and that use up a lot of paper in one shot. The reason that phone books are so difficult to recycle is that the grade of paper is extremely low (lower than newspaper) and has a high ink content, the dye used on the paper is extremely tenacious, and the bindings use so much glue. In the recycling plant this glue presents a problem by sticking to moving parts or by plugging drainage holes. Thus, it is necessary to find a processor who will either cut off the bindings or who has a process by which the hot glue can be managed.

Some uses for recycled phone books include animal bedding, tissue paper, recycled paper board, ceiling tile, drywall, hydroseed, packing paper, newsprint, insulating materials, construction paper, and molded pulp.

Unacceptable Materials in Telephone Book Recycling

- *Phone book wrappers:* The clear or colored plastic wrapping on new phone books must be removed for the book to be recycled.

Glass

- The average American uses close to 200 glass containers per year.
- A glass bottle takes one million years to decompose.
- Recycling a glass jar saves enough energy to light a 100W bulb for four hours.

Glass is 100% recyclable. Old glass containers can be made into completely new glass containers with no loss of quality and a very small loss of materials. It is a closed-loop commodity that may cycle back to consumer form innumerable times. Glass purchased from suppliers and processors can also be used in the production of fiberglass and in road construction in an asphalt substitute known as glassphalt.

Unacceptable Materials in Glass Recycling

- *Food matter:* Jars must be completely empty before recycling.
- *Drinking glasses, window and mirror glass, and crystal:* Non-used beverage container (UBC) glass is made from tempered glass. The tempering and strengthening of glass is achieved by using lead in the manufacturing process. Crystal also contains certain quantities of lead (hence the term "lead crystal"). When these materials are melted down, the two by-products are glass and lead. Obviously, the glass is good. Unfortunately, current recycling plants do not have any way to strain or siphon off the lead by-product, so the plant must shut down and physically scrape off the lead. Although extremely costly, if the time is not taken to remove the lead it will superheat and eventually burn through the cauldron, forcing the plant to shut down for even more costly repairs. No lead means no problem, so always use unleaded glass if possible.
- *Ceramics and heat-resistant glass (cookware, laboratory glass, etc.):* Although heat-resistant glass certainly is glass, and ceramics while not actually glass have many glass-like properties, neither of these products melt at the temperature used in a glass container furnace (their melting point is much higher). The nonmeltables will show up in other glass products as "stones" or other defects which, although perhaps pretty to bottle collectors, make the product commercially unmarketable.
- *Lightbulbs and fuses:* Light bulbs and fuses are not acceptable due to the fact that the glass elements in them are tempered and also must be separated from the metal filaments and housings, a process that UBC recycling plants are not equipped for. Fluorescent lightbulb mercury recycling issues are often included in the institution's hazardous-waste-planning programs.

Plastics

- The average American uses the equivalent of 32 two-liter plastic soda bottles per year.
- The plastic from five soda bottles can completely insulate a ski jacket.
- If 15 people recycled all their milk jugs for a year, together they could build a six-foot plastic park bench.

Recycled plastic can be put to a number of uses. Depending on the type of plastic, it can be remade into new plastic containers, plastic bags, packing or insulating materials, carpet fibers, parking curbs, speed bumps, plastic lumber, flower pots, buckets, trash cans, and many other things.

There are many different types of plastics and resins, not including the blends that some manufacturers use. This makes a systematic recycling of the different grades very difficult. In addition, some plastics manufacturers put cornstarch into their product to make it "biodegradable" (plastics are not biodegradable; they will only break down into smaller pieces of plastic rather than their component molecules), but this makes it nonrecyclable, as the cornstarch will contaminate the other plastics put in for recycling and lower the quality of the product.

Unacceptable Contaminants in Plastics Recycling

- *Plastic cutlery:* Disposable plastic cutlery is not normally made of recyclable materials but of other resins (usually polystyrene). Some more-permanent plastic kitchenware is recyclable, but it's best to check.
- *Plastic wrap and grocery bags:* Plastic films (bags, photographic film, plastic packaging, etc.) are not recyclable due to the wide variety of plastics that are used in their manufacture and the fact that plastic grocery bags are made through a different process from other containers. The best solution is to use them as garbage bags for other nonrecyclables (blue plastic bags are for recyclables) or to reuse them the next time you go shopping.
- *Floppy discs:* Computer discs are not made out of the same kind of plastic as most plastic bottles, and they also have metal and magnetic components.

Steel

- Recycled steel is chemically identical to steel made from virgin ore.
- The energy saved from recycling steel each year is enough to power the city of Los Angeles for eight years.
- The United States scrap industry recycles 56 million tons of steel each year—only 40% of maximum capacity.
- The average American uses 142 steel cans per year.

The popular expression "tin can" is a gross misnomer. Tin cans are actually steel cans, but in the past they were plated with tin to prevent corrosion. Modern methods rarely call for tin in the production of steel cans, but, however they are made, they are highly recyclable. The EPA lists seven major benefits of recycling scrap steel and iron into new products rather than using virgin materials. Recycling steel uses 74% less energy, 90% less virgin materials, 40% less water, creates 86% less air pollution, 76% less water pollution, 97% less mining wastes, and a whopping 105% less consumer waste. For every ton of steel recycled, 40 pounds (18.4 kilograms) of limestone, 1,000 pounds (453 kilograms) of coal, and 2500 pounds (1130 kilograms) of iron ore are saved. If we recycled steel at full capacity, each year we would save enough energy to power Los Angeles electric needs for 20 years.

Steel is America's most recycled resource, and it is not affected by the same contaminants that usually affect other recyclable materials. The amount of tin in steel cans today is too little to create problems in the furnace, and aluminum is actually necessary in the recycling process in order to increase the furnace's heat.

Steel is to be recycled in the same containers as glass and plastic. As with other food containers, steel cans must be rinsed out first and then disposed of. It is not always necessary to remove the label, but it is very helpful to do so. Paint cans and aerosol cans are also acceptable, but as with any other recyclable, they *must* be *completely* empty for disposal.

Batteries

Batteries are recyclable, but with specialized processors only, because it is especially important that the facilities are able to deal with the acid from the batteries. They should not be recycled with steel. The battery acid would change the chemical composition of the steel, and batteries also tend to explode when cooked in a blast furnace in a scrap smelter.

Unacceptable Contaminants in Steel Recycling

- *Bottle and jar caps:* because of the plastic and rubber rings on the inside used for hermetic sealing

Aluminum

- Every three months Americans throw away enough aluminum to rebuild every plane in our commercial air fleets.
- The average American uses over 360 aluminum cans per year.
- An aluminum can takes 500 years to decompose.
- Recycling aluminum uses 95% less energy than producing aluminum from raw materials.

Aluminum has the highest value of all of our recyclable materials—in fact, when it was first discovered in the 1820s, it was worth more than gold! Napoleon Bonaparte even had a dining set made out of solid aluminum to show off his wealth. Aluminum cans are highly recyclable and lose only 5–15% of their material during the recycling process. The energy conserved by recycling is so great that for every pound of aluminum (28.87 12-ounce cans) recycled, 7.5 kilowatt hours of electricity (enough to run a 100W bulb for 75 hours) can be saved. In addition, the process of recycling aluminum is a very expedient one. In as little as six weeks, an aluminum can may be manufactured, used, and recycled back into a new can again.

Unacceptable Materials in Aluminum Recycling

- *Any nonaluminum product:* Because of the extremely high value of aluminum relative to other recyclables, it is very important to keep it separate from other materials which act as contaminants in the recycling process and thus drastically lower its value to the processor.
- *Food matter:* Pour out any flat, leftover soda and rinse out the can before recycling.
- *Aluminum foil:* Although foil products are in fact made of aluminum, there are two difficulties that processors have with them. The first is that they are more than likely to be contaminated with food materials, making them difficult to recycle through current processes. The second is that these products are made of a different alloy, thickness, and shape of aluminum than the standard drinking can (current machinery is set up to accept only the form of the aluminum can).
- *Aluminum pie tins:* Pie tins are recyclable (their alloy is practically identical to that of the aluminum can), but they must be as close to surgically clean as possible with no residue on them. Wash these, crush them, and put them in with the aluminum cans.
- *Steel food and beverage cans and paint and aerosol cans:* Do not assume that all metals are created equal. Steel and tin are recyclable, but they go in receptacles separate from aluminum cans.

Aluminum cans should be completely empty and rinsed out before recycling. Concentrated moisture in a recycling furnace can cause a steam explosion.

10.4   IDENTIFICATION SYMBOLS FOR RECYCLABLE PRODUCTS

Coding of container ingredients should provide useful information with which to plan an appropriate recycling system for the institution's collected and separated waste materials, for example, glass jars. Beware of a common misunderstanding:

This is an identification of material types and *not* an assurance of feasible recyclability. Just because a container has the familiar triangle of chasing arrows does not mean that it is made of recycled materials. Product labels should specify in writing the recycled content. If a product is only identified as "recyclable," this does not mean that it is not recycled as well.

10.5   THE EPA WASTE HIERARCHY: *REDUCE, REUSE,* AND *RECYCLE.*

Source Reduction of Waste

By reducing the amount of nonrecyclable waste we create, we can reduce the rate of landfill expansions. In order to cut down our waste output, we must cut down on our waste input. Buy products that do not have excessive packaging. Compact-disk packaging was an oft-cited example several years ago. Although the record industry is phasing it out, some stores independently remove the existing pressboard box, reshrinkwrap the jewel box, and sell the CD without the pressboard. While the record stores may not have had the choice of how they received the product, you do. Cut down on individually wrapped food items. Reduce your junk mail by writing to the companies that send you things you'll never read and asking them to take your name off their list and asking them not to give or sell your name to mailing-list companies. Cancel subscriptions to magazines or papers that you don't actually read. Send E-mail instead of office memos, or if you must send a hard copy, print it twice per page and send half sheets. Correct your first draft on the computer rather than printing it. Use both sides of the page when printing or photocopying. Reducing your material input lowers the amount of resources used in your institution, and consequently reduces the amount of waste that must be landfilled, incinerated, or recycled.

Reuse of Containers

Don't just buy something and throw it away. Bottles can be used as vases or for storing water, jars can be used for storing leftovers. Plastic and paper bags can be used more than once—use them until they fall apart. Better yet, use canvas shopping bags that will last indefinitely and are easy to wash. Buy rechargeable batteries. Use old newspapers or popcorn as packing material instead of foam peanuts. Use old reports or memos as scratch paper before recycling them. Take a mug to the office instead of using paper or foamed polystyrene cups. We are so used to the disposable life that plastics were supposed to create, and now we find that they aren't really so disposable. Reuse what you have.

Recycle

Once is not enough. It is not good enough to say that you will recycle when it becomes necessary—it already is. Educate yourselves. Educate your community. Educate your family. Educate your children—the next generation should never know a world without recycling. Even if there is no law in your area that mandates recycling, take the initiative and recycle on your own. Everything is recyclable. It is simply a question of having the right processor for it. The many items that are listed as nonrecyclable in this section are not *easily* recyclable. In different locations, processors of these waste streams can be found. It becomes, then, a question of economy. When the only option is to transport recyclables to an out-of-state processor that does not give a high-enough return on a load to cover the cost of delivery, it makes more sense and saves money to dump it or incinerate it. Although the point of recycling is to save resources and cut down the waste stream, nobody is going to continue recycling at a cost to themselves; recycling *is* a business.

10.6   WASTE-MANAGEMENT OPTIONS

As the chief regulatory agency responsible for protection of human health and the environment, the EPA has established methods that it recommends to reduce the unfavorable impacts of waste disposal. EPA's priorities have been ranked as follows:

1. waste reduction
2. separation and concentration
3. exchange or recovery
4. incineration or other treatment
5. security or ultimate disposal

Source Reduction

This means the reduction or elimination of hazardous waste at the source, usually within a process. Source-reduction techniques include substitution, process modifications, housekeeping- and management-practice changes, increases in the efficiency of equipment, and recycling within a process.

Recycling

Recycling applies to any practice that processes a previously used material for reuse. It includes the reclamation of useful constituent fractions within a waste material or the removal of contaminants from a waste to allow it to be reused.

Treatment

Any method, technique, or process that changes the physical, chemical, or biological character of any hazardous waste so as to neutralize such waste, to recover energy or material resources from the waste, to render such waste nonhazardous, less hazardous, safer to manage, amenable for recovery, amenable for storage, or reduced in volume.

Disposal

The discharge, deposit, dumping, injection, or placing of hazardous waste into or on any land or water so that such waste or any constituents may not enter the air or be discharged into any waters, including groundwater.

10.7   WASTE MINIMIZATION

The Dictionary of Terms in the Safety Profession (3rd edition, ASSE) defines hazardous-waste minimization, referred to as hazardous-waste reduction, as any of various activities designed to reduce or minimize production of hazardous waste at its source with the goal being to lessen all the problems associated with hazardous waste by decreasing its creation at the source. The EPA's *Waste Minimization Opportunity Assessment Manual* defines waste minimization as source reduction and recycling.

Other disposal options are limited in scope. Restrictions have created a scarcity of approved treatment, storage, and disposal facilities (TSDFs), an RCRA disposal status discussed in Chapter 8, which results in higher disposal costs. In addition, past generators of hazardous waste are today's potential responsible party (PRP) with immense legal liability and associated costs of cleaning up TSDFs that have polluted their surroundings through improper management practices and inadequate technologies. On uniform hazardous-waste manifests, large-quantity generators must certify that they have a program in place to reduce the volume and toxicity of waste generated; small-quantity generators must make a reasonable effort to minimize waste generation and select the best waste-management method available. These financial and legal incentives make establishing a waste-minimization program attractive.

10.8 METHODS FOR WASTE MINIMIZATION

Change Materials-Purchasing and -Control Methods

Usually, raw materials and supplies are bought based on minimizing purchase cost without examining disposal cost. Profits for chemical companies are based on volume sales. If only one liter of a chemical is needed for an experiment or operation, don't buy a case and save the rest for hazardous-waste disposal at a later date.

To Minimize Disposal of New or Unused Material

Reduce the number of different products (i.e., cleaning fluids, solvents, etc.) used at your facility. Buy in container sizes appropriate to the actual use. Reduce the inventory of hazardous materials, and ensure that old containers are rotated from the back of shelves to the front when new material is purchased.

Improve Housekeeping Practices

Surplus waste production often results from poor housekeeping practices. Leaking valves can cause chemicals to spill onto the floor, requiring cleanup and disposal of wastes.

Substitute Less-Toxic Materials

Usually, a nontoxic material can be substituted for one that is toxic or that causes a waste-treatment problem. The reduced cost of disposal or the reduced exposure of workers to a toxic material can validate and support the change.

Change Process Methods

There is a tendency to continue using the same processes even after enhanced methods have been developed. However, the process can be impacted in other than a production-related sense if it produces too much waste. Frequently, changes that result in waste reduction can lead to production improvements as well.

Segregate Wastes

Wastes that may require uncommon treatment or disposal should be separated from other wastes. EPA's view of a hazardous-waste mixture takes the position that generally, the mixing of a regulated hazardous waste with a nonhazardous waste renders the whole mixture legally hazardous. Therefore, segregation of wastes makes major economic savings possible. Also, recycling a material usually requires that the waste remain as clean as possible prior to reuse.

Recycle/Reuse/Reclaim

It is often less expensive to recycle a chemical than it is to purchase new material and pay for disposal costs. Sometimes a material may no longer meet the specifications for a process in which it is being used; however, the material may still be acceptable for other uses at the facility.

Treat to Reduce Volume and/or Toxicity

Disposal costs are based on the classification of wastes and their volume. If a waste has hazardous characteristics, then disposal in a hazardous-waste disposal facility is required. If the waste does not have hazardous characteristics, it can be disposed of in an industrial-waste landfill, which is considerably less expensive. Disposal costs can be minimized by reducing the volume. Use of microexperiments in a chemistry lab, or changing the experiment so that the final components are not considered a hazardous waste are two simple ways to reduce volumes.

Work to "Delist" Wastes That Are Not Toxic

A waste can be "listed," or regulated, as hazardous because it comes from a process that has generally been found to produce hazardous wastes. This regulation may compel an instititution to dispose of a waste in a hazardous-waste disposal facility even though the waste does not actually have hazardous characteristics. There is a regulatory procedure by which a listed waste can be delisted if it can be established that the waste does not exhibit hazardous characteristics. Since the delisting procedure can be expensive and time-consuming, it should be used only when the cost savings would be significant. Coalitions and associations typically take the lead in this expensive process. Delisting effort will usually not be a good option for savings for an institution.

10.9   ELEMENTS OF WASTE MINIMIZATION

Waste minimization should become part of the insititution's "Administrative Mandate" to all employees and residents. In today's environment of business reengineering, each money-saving operation will contribute to the overall health of the institution. If the institution is to encourage employees and staff members to begin and sustain waste-reduction changes as part of their daily duties, each individual needs to realize that minimization is an important goal of the health-care or educational institution. It is the "right thing to do" for the modern institution. The management commitment to waste minimization is easy to state but the commitment has to be real, has to affect operating decisions, and has to be a part of job-performance evaluations for the managers, faculty, and individual staff members. Administration policy and programming should ensure that waste minimization is an institution-wide effort.

Characterization of Waste Generation

Maintain a waste-accounting system to track the types, amounts, and hazardous constituents of wastes and the dates they are generated. Use periodic waste-minimization assessments as a tool to measure progress. Track materials that ultimately wind up as waste, from the entry into the institution at loading dock to the point at which they become a waste.

- Determine the true costs of the waste.
- Identify opportunities at all points in a process where materials can be prevented from becoming a waste.

A Cost-Allocation System

Departments and managers can be charged waste-management costs for the wastes they generate, factoring in liability, compliance, and oversight costs. If an institution cannot charge back because of administrative reasons, the figures can still be made available and publicized. By doing this, an increased awareness level should result in savings.

Technology Transfer

Seek or exchange technical information on waste minimization from other parts of your institution, from other institutions, trade associations, state technical-assistance programs, or professional consultants. Use the faculty and research managers as sources of new cross-program savings ideas.

Appoint a Waste-Minimization Coordinator

Depending on the size and scope of the institutional operation this individual could have other responsibilities and be in any department. The waste-minimization coordinator should report to someone who has responsibility for both environmental and regulatory compliance. The coordinator should have training in environmental laws and issues. Also, he/she should have a basic understanding of all processes within the institution.

Make Waste Minimization a Part of the Big Picture

Individuals who are responsible for contributing to the hazardous-waste production at an institution should be personally evaluated on strategies that reduce waste-disposal costs. Also, many employees are motivated by a yearning not to pollute the areas and surrounding community in which they work. Students and faculty alike should be committed to this goal.

Program Evaluation

Conduct a periodic review of program effectiveness. Provide rewards and recognition for successful minimization efforts.

An incentive program for submitting ideas is useful only if the "good ideas" for potential waste-minimization projects are successfully implemented and maintained. In addition to rewarding those who originate the ideas, an effective program also provides rewards and recognition for individuals who do the hard work of implementing projects and ensuring their success. Incentive programs should be visible, related to performance objective and goals, flexible enough to adapt to changes, cost efficient, and above all, valued.

10.10 HAZARDOUS-WASTE-REDUCTION TECHNIQUES

Inventory Management

Proper management of raw materials, final products, and the associated waste streams is now being recognized by institutions as an important waste-reduction technique. In many cases waste is just past-expiration dated, contaminated, or unnecessary raw materials, spill residues, or damaged final products. The cost of disposing of these materials not only includes the actual disposal costs but also the cost of the lost raw materials or product. This can represent a large economic impact on any institution.

Inventory Control

Approaches for controlling inventory range from a basic change in ordering procedures to the implementation of just-in-time use techniques. Most of these techniques are well known. However, their use as very effective waste-reduction techniques has not been widely recognized. Many institutions can help reduce their waste generation by tightening up and expanding current inventory-control programs.

Material Control

An important area frequently disregarded or not given enough focus in many institutional facilities is material-control procedures. Proper material-control procedures will ensure that materials will be used without loss through spills, leaks, or contamination. It also will ensure that the material is efficiently handled and used in the course of doing business and does not become waste. An institution needs to keep track of the parts of the big picture: A retiring professor or teacher who has a 30-year-old lab needs to provide advance notice of the lab closing so that proper identification of chemical containers can be done while the knowledgeable individual is still around. Administration should keep those with hazardous-waste-disposal responsibilities informed of retirements, transfers, or terminations of individuals who have potentially hazardous waste under their jurisdiction.

Operational Modification

Enhancing the efficiency of an operation can minimize waste generation. Use of this technique can help reduce waste at the source of generation, thus decreasing waste-management liability and costs. Some of the most cost-effective reduction techniques are included in this category; many are simple and relatively inexpensive changes to production procedures. Available techniques range from the use of microscale laboratory experiments to installing state-of-the-art equipment that minimizes the use of chemicals. The waste-reduction techniques in this category can further be divided into improved maintenance, equipment modifications, and material change.

Volume Reduction

Volume reduction includes techniques to separate wastes and recoverable wastes from the entire waste stream. These techniques are usually used to maximize recoverability or decrease the volume and thus the disposal costs. The various techniques used range from simple segregation of wastes at the source to complex concentration technology. They are applicable to all types of waste streams. These techniques can be divided into two general areas: waste concentration and source segregation.

Recovery

Recovering wastes can provide a cost-effective waste-management option. This technique can help eliminate waste-disposal costs, reduce raw material costs, and possibly provide income by reducing costs of new materials. Recovery of wastes is a widely used practice in many processes and can be done on-site.

10.11 WASTE-MINIMIZATION ASSESSMENT

A waste-minimization assessment is a detailed procedure with the objective of outlining and identifying ways to minimize or eliminate waste. The procedure is a management tool that assists in reviewing an institution's processes and waste-streams. It is a major element of a waste-minimization program. The steps involved in conducting a waste-minimization assessment are outlined in Table 8-3.

The waste-minimization program helps reduce disposal costs and permitting requirements; improves public image; reduces liability under OSHA and the Comprehensive Environmental Response, Compensation, and Liability Act (CERCLA); and leads to savings in operating costs and raw-material use. Despite these advantages, the waste-minimization committee/waste-minimization coordinator must be aware of obstacles to establishing a program. These include technical and economical considerations as well as regulatory and behavioral issues.

While large institutions can provide a sizable budget to develop waste-minimization programs, smaller institutions may be unable to make the investment and commitment to successfully institute such a program. Most managers observe the bottom line, and in the process, costs associated with waste minimization are not realized. Indirect benefits, such as generating positive public images, are difficult to evaluate. Profitability and decreased operating costs are the major economic evaluators. While capital costs (e.g., equipment purchases, permitting costs) must be considered, savings in operating costs (e.g., reduced waste-disposal costs, and insurance and liability savings) should be favorable to management.

Technical considerations focus on the potential success of certain applications in a given environment and/or operation. Some changes may need to be tested on a small scale, over a certain period of time. An institution's physical constraints must also be carefully considered. Other areas to be addressed include worker safety, space limitations, and probability for causing other environmental problems. The waste-minimization coordinator and some committee members should be aware of the latest technology that applies to the institution's operations. The knowledge helps determine the feasibility of equipment and layout changes.

Another technical consideration is logistics. A small automotive supplier implemented a waste-minimization strategy to accumulate enough waste solvent to make recycling economically attractive to management. The solvent-recycling facility, located one mile from the plant, needed at least ten drums for batch processing; the supplier collected ten drums of solvent during an 18-month period. The company then learned during a state EPA inspection that it would be fined for failure to adhere to regulations requiring that waste be shipped off-site every 90 days if a storage permit was not obtained.

Behavioral obstacles to program implementation include institutional reluctance to change existing operational processes that have been in place for years. Institutions may also hesitate to provide governmental agencies with information that might eventually be used against them.

The final obstacle is regulatory. The committee should remember that instituting a waste-minimization program can result in revising existing environmental permits at the facility.

After all possible obstacles have been addressed, it may be necessary to reevaluate which waste-minimization method(s) will be used.

10.12   SELECTING A HAZARDOUS-WASTE CONTRACTOR

If you still have hazardous waste after you have completed your waste-minimization efforts then you usually have to select a contractor who will help you with disposal of your waste.

Selecting a good hazardous-waste contractor is very important. Since most institutions use contractors for at least some part of hazardous-waste-disposal operations, it is very important to develop a good evaluation scheme and follow through when evaluating a contractor.

A pre-selection assessment should examine the following items:

Contract Language

All provisions of all duties and responsibilities for both the contractor and the institution should be carefully spelled out in the contract. The contract should contain information on insurance requirements and prices for the various hazardous-waste operations that will be performed. Many times the contractor with the low bid will have the following clause or something similar in the cost list: "Other required operations will be quoted on an as needed basis." This can very easily make the low-bid contractor the one who presents you with the highest bill. Your legal advisor should closely examine the contract.

Original insurance certificates should be received before any work commences. The insurance company should be contacted as to the current disposition of the policy(s).

Résumés of individuals who will be working on-site should be reviewed before the contractor arrives on site. Copies of training certificates should be included in the résumé package. You should check to make sure résumés match the workers. One contractor sent an institution a very nice package of résumés and training certificates. There was a bit of a problem when the university manager found that he could not match any of the workers on-site to any of the résumés or training certificates. If a contractor will be on-site for more than one pickup, specify in express written contract terms that you want the same workers on-site for each pickup. Many times a contractor will bring in the "first team" on the initial pickup and then start shuffling new people in on future pickups. This can be prevented by specifying in the contract that you want the same individuals on-site for each pickup and by maintaining résumés on file.

References should always be checked. Each contractor always has a ready list of references that can be checked. I have never received a list of references that were not completely satisfied with their contractor.

A list of the hazardous-waste generators in your area can be obtained from state officials, if the state operates its own RCRA program, under the state equivalents of the federal Freedom of Information Act (FOIA). Your legal counsel should assist in making these requests, so that timely responses can be obtained. If you think that an FOIA request is too much trouble, it is very easy to call other local institutions to see who they are using. Due to the nature of the hazardous-waste business, it is very likely that the contractor knocking on your door is doing business somewhere else locally. With just a few phone calls to the other local generators, a more realistic picture of the level of services can be obtained than from the carefully screened reference list that is supplied to you in the standard bid package.

Local and Federal Occupational and Safety and Health Administration (OSHA) regulatory authorities should be checked to determine if the contractor is being inspected in response to employee fatalities or multiple employee hospitalizations. The Environmental Protection Agency (EPA) should be checked to determine if the contractor has any violations for past practices due to noncompliance with applicable state and federal regulations.

Transportation is an area that can get your institution into trouble very quickly. Typically, at institutions, a "waste packaging for transportation" (labpack job) is conducted during a regular workday, eight to four or nine to five. If waste is not left on-site at the end of the day, many contractors will take your waste to a transfer station or storage facility to wait for approvals for the disposal of your waste.

Several years ago a transportation-related hazardous-material accident occurred on the beltway that surrounds Washington, DC. The accident occurred during rush hour and traffic was stopped for approximately six hours. The local response was tremendous. News reporters and hazardous-material teams responded from all over. The incident was front-page news and the lead-off story on all the local television and radio stations. Such front-page news could have been making my institution (and maybe me) very unpopular with people—not to mention the potential for injury, liability, and cleanup costs associated with a hazardous-material accident during rush hour. Transporting institutional hazardous waste deserves your individual attention.

At Georgetown University, our office immediately instituted a policy of holding hazardous-waste shipments before they leave the site until either Saturday or Sunday mornings. The first time I had a pickup using this policy, I identified another potential problem. The driver was quite proud of the fact that he had driven through the night to reach Washington by the scheduled pickup time of 7 A.M. on Saturday morning. I realized that I was getting ready to send out a full shipment of hazardous waste with a driver who had only slept two hours in the previous 24. Now, if the driver is coming from more than a few hours away, I require him to come into town on the previous night and stay in a hotel. The costs of weekend transportation are a little higher than weekday shipments but a few hours of overtime and the cost of a hotel room is a small price to pay for reducing the potential transportation liability.

Treatment, storage, and disposal facilities (TSDF) should be inspected before use. A sample contract is found in the Appendix following this chapter. I receive many brochures from TSDFs with pictures of wooded forest glens or beach-front ocean views. I have yet to inspect a TSDF in a wooded glen or with an ocean view.

Basic cleanliness of the site is a good indicator of whether good work practices are followed at the TSDF. The inspection should be documented and should cover not only a physical visual evaluation but a check of record-keeping practices. Again, the FOIA can be used to determine if there are any regulatory problems with the facility. Since many institutions cannot afford to send someone to all the TSDFs in the area, it is a good idea to exchange information with other institutions in the area. People in the business of health and safety are usually very happy to share information. Also, by communicating with others, you can develop a scheme whereby you can take turns or make sure you are evaluating different TSDFs. If an on-site inspection is not possible, a checklist can be sent to the facility to gather information.

When the contractor comes on-site, several items should be evaluated. There should be a site-specific safety plan. All workers should be aware of the components of that plan. You should quiz the workers on what procedures would be if there was an emergency. The inventory of on-site equipment should be checked. Does each worker have access to personal protective equipment (PPE)? Was spill-response equipment brought on-site? Do the contractors have set protocols for dealing with high-hazard material? Do the contractors follow their own protocols?

Paperwork trails should always be carefully maintained. Hazardous-waste manifests should be sent to their respective destinations by registered mail. Return receipts should be kept with the paperwork file. A checklist in the form of a time line should be maintained until all respective paperwork has reached its final destination. If you have not received a signed copy of your manifest from the transporter within 35 days, you need to contact the transporter to determine what has happened. The EPA regional administrator must be contacted in writing if you have not received a manifest within 45 days. The letter must contain a copy of the manifest in question and detail the efforts taken by the generator to locate the hazardous waste and be signed by the generator's responsible party.

It is a good idea to document the evaluation of each contractor for each pickup. There are many forms that the evaluation can take. One easy method is to use a "report card" system for each pickup or emergency-response situation that deals with hazardous waste. The contractor can be evaluated in ten different categories with a one-to-ten score. The categories are:

- insurance
- prices
- TSD facilities
- site safety plan/emergency preparedness
- résumés/worker's level of knowledge
- professionalism
- safety equipment
- protocols
- manifests
- support documentation

There is no failing or passing grade because the criteria is not being weighted, but at least there can be a relative comparison of different contractors.

Whatever evaluation system you use, maintain a consistent program and document your results.

The proper management and disposal of chemical hazardous waste is essential to the health and safety of institutional personnel as well as to individuals in the surrounding community.

10.13 SUMMARY ON WASTE MANAGEMENT

The mismanagement of waste can become a very serious institutional environmental problem. Generators of waste must continually seek ways to improve management techniques and explore new options for cost-efficient management. Reduction of waste by elimination and minimization should be a primary institutional goal.

REFERENCES

Abercrombie, Stanley A. 1988. *Dictionary of Terms Used in the Safety Profession.* 3d ed. Des Plaines, IL: ASSE.
Code of Federal Regulations Environmental Protection Agency, 40 CFR, Part 261.3, Office of the Federal Register, National Archives and Records Administration, Washington, DC, July 1989 edition, p. 145.
EPA Hazardous Waste Engineering Research Laboratory. 1988. *Waste Minimization Opportunity Assessment Manual.* Cincinnati: EPA.
Miller, Jon. 1995. *Georgetown University Recycling Manual.* Washington DC: Georgetown University.
O'Reilly, James T. 1992. *State and Local Government Solid Waste Management.* Deerfield IL: Clark Boardman Callaghan.
O'Reilly, James T., Kyle Kane, and Mark Norman. 1993. *RCRA and Superfund Practice Guide.* 2d ed. Colorado Springs: McGraw-Hill/Shepards.

Suggested Hazardous Waste Contract

The following is a sample institutional contract that may offer some valuable ideas as your own institution prepares to deal with its hazardous-waste contractors. Any such document must be reviewed by the institution's own legal counsel prior to use.

AGREEMENT

THIS AGREEMENT made this (number) day of (month), 19__ by and between (Contractor), (Address) (hereinafter called "(Contractor)") and Institution, (Address), (hereinafter called "(Institution)").

WHEREAS, "(Contractor)," a (Name of Applicable Jurisdiction) business entity that contracts with other entities engaged in the business of handling, packaging, labeling, removal, transportation, treatment, and disposal of hazardous chemical waste; and

WHEREAS, "(Institution)" is desirous of contracting with "(Contractor)" upon specific terms and conditions for assistance in the packaging and removal of certain hazardous waste.

NOW, THEREFORE, in consideration of the following promises and premises set out herein, and for other good and valuable consideration, the parties agree as follows:

1. EFFECTIVE DATE OF CONTRACT

The terms of this contract will commence on (DATE) and terminate on (DATE) . The termination date notwithstanding, this contract shall remain in full force and effect until such time that the parties have fulfilled all of their respective obligations hereunder, including, but not limited to the submission of disposal-activity documentation that may be due to "(Institution)" or any regulatory body. This contract may be renewed thereafter upon mutual agreement of the parties for a like term.

This contract may be canceled by "(Institution)" at any time during the term hereof only if "(Contractor)," and/or its agents fail(s) to deliver promptly, fails to provide responsible service consistent with recognized industry standards, or fails to meet the terms and conditions of the contract.

This contract may be terminated by "(Contractor)" at any time during the term hereof only if "(Institution)" fails to make prompt payment as more fully set forth in paragraphs 3 and 4 herein or other terms of this contract.

2. REMOVAL, TRANSPORTATION, AND DISPOSAL OF HAZARDOUS CHEMICAL WASTE

A. "(Contractor)" and/or its agent hereby agree to remove "Institution" 's chemical waste for off-site disposal. "(Institution)" will provide a repre-sentative list of all hazardous waste intended for off-site disposal to "(Contractor)" prior to the scheduled pickup date. "(Institution)" Department of Safety, will assemble all waste intended for off-site dis-posal and turn over all said waste to "(Contractor)" and/or its agents at a predetermined pickup area on "(Institution)" premises. "(Contractor)" and/or its agents will prepare and perform lab packing and other func-tions necessary to transport the waste from "(Institution)" premises according to applicable rules and regulations of the Environmental Protection Agency (EPA) and the Department of Transportation (DOT) and/or any other regulatory agency having jurisdiction over said opera-tions.

B. The first pickup of waste may be required by "(Institution)" within (30 or some other number) days after this contract is signed and may there-after be performed at the request of "(Institution)" on a regular basis to a disposal facility that utilizes incineration and/or recovery/chemical treatment as the ultimate disposal method, and that is permitted by the EPA and local authorities for said hazardous-waste-disposal activity. After either of these two disposal methods are completed, "(Contractor)" and/or its agents agree, if requested, to complete a cer-tificate of disposal duly authorized and signed by a principal of the dis-posal facility, to "(Institution)" for all such waste. For certain limited waste which cannot be appropriately disposed of by incineration and/or recovery/chemical treatment and where a certificate of disposal is unob-tainable, "(Contractor)" and/or its agents must notify "(Institution)" and recommend an ultimate disposal method. "(Institution)" will require cer-tification from "(Contractor)" to the effect that ultimate disposal was accomplished according to applicable EPA and DOT laws and regulations. Any alternate form of certification shall be acceptable to "(Institution)" prior to performance of the alternative-disposal process at the time "(Contractor)" recommends the alternative method. The "(Institution)" will be notified in writing of the type of ultimate disposal used for each component of each waste stream.

C. "(Contractor)" agrees that it will immediately advise "(Institution)" Department of Safety, of any and all untoward incidents (spills, leaks, releases, notices of violations, and notices of investigations) involving "(Institution)" waste materials as soon as "(Contractor)" becomes aware of them.

D. "(Contractor)" shall prepare and execute the hazardous-waste manifest in accordance with applicable federal and/or state rules and regulations and shall provide "(Institution)" with the required number of copies.

3. COST

The parties hereby agree that the total cost for the removal, transportation, and disposal of "(Institution)" waste will be as stated in the attached price list, and shall not exceed such costs unless by prior agreement of the parties.

4. PAYMENT TERMS

"(Institution)" agrees to pay "(Contractor)" ultimate disposal costs described in section 3 above within thirty (30) days following receipt of "(Contractor)" invoice after disposal has occurred. "(Institution)" agrees to pay "(Contractor)" all associated costs other than DISPOSAL cost within thirty (30) calendar days following receipt of "(Contractor)" invoice.

5. "(CONTRACTOR)" WARRANTIES

"(Contractor)" warrants and represents that it (a) will engage the necessary entities experienced in the business of the handling and disposal of hazardous-waste material; (b) that "(Contractor)" is aware of the hazards associated with such activities and is proficient in the procedures necessary to deal with them in a manner consistent with recognized industry standards; (c) that "(Contractor)" will comply with all applicable laws and regulations relevant to any activity to be performed pursuant to the contract; (d) observing all appropriate safety precautions; (e) that "(Contractor)" and/or its agents will obtain, and will maintain throughout the Contract, all necessary government authorizations for "(Contractor)" and its agents and any specific facilities, practices, or pieces of equipment that will be used in the provision of the services; and (f) that "(Contractor)" is not currently under investigation or subject to any enforcement actions by any third parties in any of the jurisdictions where the contractor is performing business (the contractor will immediately notify the Institution if such investigation or enforcement actions occur); and (g) shall comply with all current federal, state, and local laws, regulations, and ordinances applicable to the transportation, treatment, storage, and disposal of hazardous waste.

"(Contractor)" shall obtain all permits, licenses, and current documentation required in order to comply with such laws, regulations, and ordinances and shall assure that all subcontractors of "(Contractor)" will also obtain all necessary permits, licenses, and any other required forms of documentation. "(Institution)" will receive either original or copies of all the aforementioned documentation prior to commencement of work.

The contractor will notify in writing the "(Institution)" if any of the above conditions occur during the life of the contract.

6.   TRANSFER OF TITLE, RESPONSIBILITY, AND INDEMNITY BY CONTRACTOR

Following loading and departure from "(Institution)" premises "(Institution)" shall be relieved of responsibility and "(Contractor)" and/or its agents or affiliates shall become responsible for any and all loss, damage, or injury to persons or property and "(Contractor)" shall indemnify and hold "(Institution)" harmless from any and all liability, fines, damages, costs, claims, demands, and expenses of whatever type or nature, including, but not limited to, pollution or other damage, which shall be caused by, arise out of, or in any manner be connected with the waste-disposal services, "(Contractor)" shall also indemnify and hold "(Institution)" harmless from any and all liability, damages, costs, claims, demands, and expenses of whatever type or nature resulting from the acts, errors, or omissions of "(Contractor)", its employees, agents and subcontractors. "(Contractor)" and/or its agents or affiliates will assume Title and Total Responsibility for the waste materials at the point in time when the waste materials are loaded on the transportation vehicle at the designated "(Institution)" pickup point.

7.   "(INSTITUTION)" RIGHT TO INSPECT

"(Contractor)" agrees to allow "(Institution)" to inspect and obtain copies of all licenses, permits, or approvals required for the performance of waste-disposal services issued to "(Contractor)" and/or its agents, affiliates, or subcontractors by any governmental or regulatory entity or agency.

"(Contractor)" shall also permit "(Institution)" to inspect transportation vehicles and containers, and to inspect the transportation, unloading, storage, treatment, and disposal operations conducted by "(Contractor)," in the performance of this contract. Although "(Institution)" has the right to inspect, "(Institution)" is not obligated to do so and "(Contractor)" is not relieved of its own responsibilities under the Contract or the law by this right to inspect.

8. "*(INSTITUTION)" FACILITY—ACCESS AND EMERGENCY PROCEDURES*

A. "(Contractor)" employees, agents, and subcontractors shall enter "(Institution)" premises only with the permission of "(Institution)" Department of Safety, Medical Center, and will be accompanied by an authorized representative of "Institution" Department of Safety.

B. Emergency Procedures—In the event an emergency occurs while a "(Contractor)" employee, agent, or subcontractor is on the premises of "(Institution)," "(Institution)" shall make available to such person or persons emergency, including first aid, to the same extent that emergency services would be available to an employee of "(Institution)" on the same premises. While the waste-disposal services are being performed on "(Institution)" premises "Contractor" will comply with "(Institution)" internal safety procedures.

9. *INSURANCE*

"(Contractor)" shall not begin any operations under this Agreement nor shall it allow any subcontractor to begin any operations until: (a) it has obtained all the insurance required herein; and (b) it has furnished original certificates of insurance to "(Institution)." Every certificate of insurance providing the coverage required herein shall contain the following clause: "No reduction, cancellation, or expiration of the policy shall become effective until '(Contractor)' has completed all waste-disposal services specified in this Contract to the satisfaction of '(Institution).' "

The "(Institution)" shall be a named insured on the policy at the expense of the "Contractor."

"(Contractor)" and/or its subcontractors shall take out and maintain for the life of this Agreement (at its own expense unless otherwise specifically set forth) at least the following insurance.

Coverage	Limits
Workers' Compensation	Statutory
Employer's Liability	$1,000,000 each occurrence
Public Liability (BI & PD)	$1,000,000 combined single limit
Automobile Liability (BI & PD)	$1,000,000 combined single limit
Pollution Liability	$1,000,000 each occurrence

The public-liability insurance shall include coverage for all of "(Contractor)" contractual liability under the Agreement with limits of not less than those set forth above. "(Institution)" agrees, however, that such public-liability insurance need not cover losses, damages, costs, or expenses arising out of bodily injury (including death) to any person or damage to any property caused by or resulting from acts or omissions of "(Institution)," its employees, or its agents.

10. ASSIGNMENT

Neither party shall assign, sublet, transfer, nor convey this Agreement or any monies due or to become due to it hereunder without the prior written consent of the other.

11. CONFIDENTIAL INFORMATION

"(Contractor)" and "(Institution)" agree not to disclose any confidential or proprietary business information of the other acquired during the course of performing this contract. For the purposes of this contract, business information will be considered confidential or proprietary if labeled as such or if the disclosing party specifically requests information to be so classified. To the extent that information classified as confidential or proprietary is required by law to be disclosed, the parties shall not be bound to any pledges of confidentiality made pursuant to this section.

12. GOVERNING LAW

The validity, interpretation, and performance of this contract shall be governed and construed in accordance with the laws of "applicable jurisdiction."

13. ENTIRE CONTRACT

This contract represents the understanding and agreement between "(Contractor)" and "(Institution)" relating to the transportation, treatment, storage, and disposal of hazardous waste. Accordingly, this contract supersedes and renders null and void all prior agreements and representations, whether written or oral, that may exist between the parties regarding the subject matter herein.

No changes, alteration, or modifications to this contract will be effective unless in writing and signed by duly authorized representatives of "(Contractor)" and "(Institution)."

IN WITNESS WHEREOF, the parties have caused this Contract to be executed by their duly authorized representatives, effective as of the below date.

_____   _____ (SEAL)
Witness
Date _____

_____   _____ (SEAL)
Witness
Date _____

Chapter 11

Grounds and Pedestrian Safety

11.1   GROUNDS AND CHEMICAL EXPOSURES

Maintenance of green space, especially turf fields for athletics, depends on care and patience. When these are in short supply or the green space is in high demand, chemical fertilizers and pesticides are sometimes substituted. The ideal is an integrated pest-management (IPM) approach with adequate watering and time to grow.

Pesticides are safe for humans and the environment when used appropriately. The Environmental Protection Agency (EPA) enforces the proper use of pesticides under authority of the 1972 Federal Insecticide, Fungicide, and Rodenticide Act (FIFRA) as amended. FIFRA provides that for the most serious health risks, only those individuals who have been specially trained will be able to purchase, use, and apply the chemicals. These persons take a test and become "certified applicators"; it is illegal for persons other than certified applicators (and those supervised by them) to handle the most serious pesticides. Examples would include a strong herbicide that is designed to clear out tangled brush, but which has severe water-contamination properties that need to be avoided.

The federal-state role in pesticides is intended to be a positive sharing of enforcement authority with limited scope for local permission in the use of unapproved chemicals. Central federal approval of new pesticides (and of old pesticides in new crops or new uses) is paramount, but the actual enforcement is done by state agricultural officials and to a much lesser extent by U.S. EPA pesticide inspectors.

The institution's use of pesticides, for example, for swimming pools (chlorine to kill germs) or stadium grass (chemicals to control fungi), should be undertaken with a well-considered pesticide-use plan that accounts for the level of risks appropriate to use, adequate time after pesticide application to minimize human contact with the crops or turf that has been treated, and a plan to clean and dispose of used pesticide containers properly. Training and education of the grounds crew in personal protective equipment, reading the labels, and exercising health-based protections will all be part of the institution's plan.

11.2 STREET AND SIDEWALK SAFETY ISSUES

The perfect amount of precipitation never seems to fall on your campus or facility; there's too much snow, too much rain, or the freeze-and-thaw cycle breeds potholes where streets have been corrupted by ice and salt over the winter season.

Street and sidewalk repairs should be properly signaled with barriers, flashers, and signs appropriate to caution cyclists, walkers, and drivers about the repair spot.

Clearing snow-laden streets with salt might not be the right environmental choice, where other forms of treatment might work just as well under local conditions. A comprehensive snow-removal plan should predict the types of removal equipment and materials that the facility needs. In this age of increased sensitivity to mobility-impaired persons, the snow-removal plan should consider egress for wheelchair users, clearing of roadways and paths that they are likely to use, and access to the labs or classrooms where the wheelchairs must go.

11.3 FLEET SAFETY

The vehicle fleet of the institution is a major source of safety concerns, especially of insurance-liability worries for the risk-management team. The institution's safety manager should be aware of the fleet manager's or vehicle maintenance manager's plans for assuring proper safety in fleet equipment. OSHA has considered but had not, as of 1996, adopted a final standard on occupational driver-safety instructions and supervision.

Elements of the vehicle-safety program should emphasize that the rules of vehicular traffic are applied the same whether the propulsion system is a conventional gas engine, an electric golf-cart utility vehicle, or a bicycle. Safety instruction and enforcement of common-sense "rules of the road" should reduce the most frequent collisions, which occur at intersections and blind driveways. Not all vehicle operators have the same sense of this basic element: Police cars can be risky at their top speed of 150 miles per hour, for example, but a campus with one mile of twisting streets and lanes is more imperiled by downhill bicyclists who routinely run stop signs.

Pedestrian-safety training should be included in student handbooks, employee introduction materials, and the like. Basic accident-prevention training is a good investment; in-line skaters and other fast-moving "pedestrians" deserve special concern for the education of these risk-prone adventurers.

11.4   CRIME PREVENTION AND PHYSICAL SAFETY

No text on grounds-safety issues would be complete without a recognition that safety includes the reduction in assaults, rapes, and other human-initiated safety problems. The balance between grounds safety and physical security brings the environmental, health, and safety team into close cooperation with campus security officials. For example, locked egress doors are a life-safety-code violation that will bring on fire inspectors' penalties. The correct way to secure against intruders without suffering risks of fire deaths can be worked through by the several professionals in the institution's management team.

Lighting and outdoor alert telephones for security contacts are an aspect to which many campus safety directors have devoted extra attention in recent years. As Chapter 2 discussed, liability to "invitees" on the grounds is a matter of significant legal debate. Proper lighting of paths and parking areas, reduction in the darker corners where loitering could lead to trouble, and ample call boxes for passersby should be part of the campus-security plan. The emergency phones must work when needed, so batteries and electrical maintenance should be checked regularly.

Vandalism is an unfortunate part of urban campus life. The safety aspect should be a standard protocol for reporting, marking of any dangerous aspect to the distressed area, and the prompt scheduling of a vandal-resistant substitute. Alcohol-based vandalism on college campuses is a particular challenge, but the quality and durability of fixtures can reduce the risk that a vandal's act will cause serious personal harm to a later passerby.

Experience with bomb threats suggests that the student population that is unprepared for an examination sometimes falls for this "way out," despite its criminal irresponsibility. One safety manager recommends that the examination announcement should routinely announce that the class or examination will be moved to another location in the event of any unforeseen emergency, but that it will be held nonetheless at the site to be announced if a move is necessary. That discourages the person foolish enough to believe that a bomb hoax will preclude taking the test.

11.5 ANIMAL-CONTROL POLICIES

Institutions that have public-health programs, liability-avoidance training, and empathy for animal rights sometimes overlook the problems that lack of adequate animal control can cause for the humans on campus. Campuses are usually subject to state and local laws regarding leash control of dogs, and campus rules should provide for adequate cleanup of the canine contributions to the fertilization of campus green space. But cats may be a larger problem.

It is estimated that the United States has between 11 million and 19 million cats that are "feral," that is, not domestic, and many of these are found in institutional settings where food can be found. The persons who feed wild cats are part of the problem; populations of feral cats expand as students or others feed the strays.

To avoid infection vectors, legal liability (human injury caused by a cat's scratching or flea-borne illness), and to promote better quality of life for the cats, some institutions favor trapping, neutering, and releasing the cats in hopes that the cat population will decline. Other groups dispute this tactic and dislike the trapping until and unless a secure, temperate climate for the release of the cats is available.

The best policy is one of banning feeding of the cats, with reminders of the ban and of the reasons for deterring further expansion of the cat population.

Mice, rats, and other rodent species may also be part of the campus environment, usually uninvited. An integrated pest-management system that accounts for safe and humane disposal is recommended. The disease potential of rat infestation is widely recognized, and the professional control techniques should include traps and removal of food sources as well as chemical means of deterrence.

Chapter *12*

What Other Regulatory Controls Apply to This Institution?

12.1 OVERVIEW

The preceding chapters have discussed the major areas of regulatory concern that will affect managers of hospitals, schools, and universities. In an effort to serve the needs of this special community, this chapter will go beyond the "big ticket" systems of regulation to advise about some of the lesser, but still important, systems of regulatory controls. The fact that a program of regulation is explored in this more-abbreviated format does not mean that the reader can take it less seriously. Attention to each of the relevant programs is necessary, and a publicized violation of an obscure rule still is a setback to the conscientious manager. The question of how to prioritize these lesser control systems is a strategic decision that must take into account the reader's particular set of needs.

Success for the individual manager may be judged by an absence of regulatory penalties, lawsuit claims, and bad publicity for the institution. But more frequently, the manager should be viewed as a winner if the well-handled response to government inspections results in learning about how to improve safety and/or regulatory agency compliance at the institution. Positive relations between the environmental manager and the relevant local, state, and federal regulators can help the institution move forward. The prudent manager is very much a learner, and government inspections are one form of the continuous, involuntary government instruction that expands the manager's view of the safety and compliance missions.

12.2   CHEMICALS—TOXIC SUBSTANCES CONTROL ACT

The federal Environmental Protection Agency (EPA) has been empowered to
impose controls on production and handling of chemicals and their components,
in the form of products and raw materials, through the 1976 Toxic Substances
Control Act (TSCA). (USC 2601) Usually, TSCA will not affect the activities of
universities, hospitals, or schools. These institutions are regulated "persons"
under the statute (In Re Villanova University 1988) but the TSCA is focused on
those in the chain of chemical production or handling, and the activities of these
institutions do not often coincide with the production-oriented actions regulated
by TSCA. Those chemicals that are regulated by the Food and Drug
Administration (FDA) as "drugs," by the Environmental Protection Agency as
finished "pesticides," or by the Nuclear Regulatory Commission (NRC) as
radioactive materials, are excluded from TSCA.

   TSCA might reach some of the activities of advanced chemical laboratories
that are affiliated with universities and industrial entities, but only if these are
sophisticated activities that are designed to produce a chemical for intentional
distribution into commercial uses. For example, a mechanical-engineering lab of
a state college may be exempt under TSCA as a noncommercial, R&D facility.
But it starts to sell the service of its special equipment for dry-powder processing
of paint pigments to aid a local tractor plant, it becomes a regulated "processor"
of a chemical mixture "for commercial purposes." The triggering events are
either the mixing or other handling of a chemical, "processing" it rather than
merely shipping unopened containers, or the importation or sale of the chemical
to others through interstate commerce.

   If the institution's laboratory or chemical-process facility is processing
chemicals and distributing them in commerce, then the institution is subject to
specific sets of TSCA reporting obligations. For example, assume that a technical-
grade polymer is purchased in bulk quantities by a college engineering department, and then is heated, distilled, and given new properties by an experimental
process that makes the polymer stronger for plastics uses. The department then
sells blocks of the plastic material to eight plastic manufacturers for $700 per
block, hoping to recoup the costs of its newest machine and perhaps to launch a
"sideline" income stream for further lab expansion. This is considered by EPA to
be both a processing of a chemical (the polymer) and distribution in commerce,
and EPA rules at 40 CFR Part 720 regarding prior EPA approval of the "new"
chemical may be triggered.

   The significance of TSCA for institutions is largely in its paperwork obligations. TSCA would be a severe inhibition to commercial development of any new
chemical substance, that is, a compound that is not on federal EPA lists of "existing chemicals." But it is reasonable to assume that the great majority of chemical
research activity of a university, hospital, or school will not seek to commercially

market the end products; education produces side benefits of patent licensing or technology know-how, but the college rarely becomes the chemical supplier for commercial producers. The most frequent response to a university discovery of a new polymer or other compound will be to license the technology to a commercial chemical manufacturer. That entity will then send the research data along to EPA when it is ready for a "premanufacture notification" in the name of the commercial entity.

TSCA is largely a paperwork-reporting burden for the affected entities. Each processor or distributor must report four principal types of TSCA information to the federal EPA, within a limited period of time:

1. any testing reports that are studies of the health and safety effects of this chemical
2. any allegations of health or environmental effects that are received from the public, for example, allegations that fumes from making the plastic are causing crop damage for neighbors' farms
3. copies or lists of any known studies that have examined the human or environmental effects of this chemical, for example, rodent-feeding studies to assess tumor potential from a new polymer
4. new information suggesting that serious harm to people or the environment may be causally related to this particular chemical

The most likely impacts of TSCA, from which the regulatory requirements will affect a nonprofit entity, are the effects on building repair and on imports. Building repair and maintenance activities may detect asbestos; leaks of fluids containing the banned chemicals called polychlorinated biphenyls (PCBs) (Villanova 1988); and lead-based paints. Institutions risk heavy fines if they do not closely follow the EPA methods for safely removing paint or insulation materials, evaluating and documenting inspections of walls, and so on. Asbestos in school buildings has been a multimillion-dollar problem for college and school administrators, because federal assistance for removal programs was significantly less in size than the federal costs imposed through TSCA standards for the removal operations.

EPA's largest nonmanufacturing TSCA penalties have been assessed for violations by entities that do not have documented programs in place for handling PCBs in their removal or storage of electrical transformers and related electrical equipment (In Re Villanova University 1988). PCB penalties are only avoided by understanding the rules, training the staff members that handle electrical equipment, and paying close attention to the EPA regulations. Consultants who have passed EPA training courses are available to guide institutions on these removal-and-repair requirements. These TSCA penalties can best be avoided by using outside consultant experts who are familiar with the PCB documentation-and-disposal requirements.

Institutions' imports of chemicals or machinery containing chemicals are also affected by TSCA, because the entity listed on Customs forms as the "importer of record" is treated by EPA as if it were a manufacturer of the chemical. When a shipment of chemicals, or of equipment with hydraulic fluid or other powder or liquid chemicals inside the equipment, is entered into the United States with a Customs form listing the hospital or university as the importing entity, the institution faces an unanticipated legal risk. Each chemical import must be individually certified to the Customs Service as complying with TSCA or as a TSCA-"exempted" shipment under 19 CFR §12.118 of Customs regulations. The penalty for filing a false Customs declaration plus the TSCA penalty for illegal "manufacture" of the imported chemical can, together, be a huge liability for the entity that imports without submitting a correct certification.

Ignorance of TSCA is not a defense, since the same item could have been imported by a competent supplier and later purchased by the institution. These TSCA regulatory requirements are well understood by many of the chemical suppliers with whom the university or hospital deals, and the institution's mistaken assumption that products are exempt is not a defense to the penalties.

Finally, TSCA allows some limited processing of chemicals under an "R&D exemption" excusing the company from the filing of an EPA premanufacture notification. This R&D exemption avoids the huge penalties that would otherwise be applied against a company that commercially distributes a "new" chemical, one that has not yet been placed on the EPA's inventory of chemical substances. Quantity and distribution of the R&D material is closely regulated by TSCA regulations. These regulations emphasize that documentation of the research purposes and technically qualified laboratory users will be vital to any TSCA exemption claims.

12.3   LEAKING TANKS—UNDERGROUND STORAGE TANK RULES

Managers of physical facilities have often learned lessons the hard way, and underground-tank leaks are among the most costly of problems to be learned. The storage of fuels, petroleum, chemicals, and other liquids in tanks and vessels that are buried underground was once a smart decision for building managers. "Out of sight, out of mind" storage of solvents, diesel fuel, and so on was formerly a wise cost-savings approach. Today's manager knows differently.

By the mid-1990s, prudent managers had eliminated as many of these hazardous-materials underground storage tanks as possible, and had replaced others with expensive new units that contain sophisticated leak-detection apparatus. The implementation of the 1984 Subtitle I amendments to the federal Resource

Conservation and Recovery Act (RCRA) put into place a set of federal norms for states to follow, under guidance of the federal EPA. EPA's rules as applied by the states will impact virtually all hospitals or schools that maintain and refuel their own vehicle fleet or that store regular or emergency supplies of liquid fuels in buried tanks (O'Reilly, Kane, and Norman 1993, Ch. 6).

Only a portion of the total population of tanks is regulated: those that contain petroleum products or chemicals that are listed as "hazardous." Typically these are gasoline refueling tanks, but many other potentially hazardous solvents and fuels are also stored and special control measures are applied to these as "hazardous" substance tanks. The new law's implementing rules in 40 CFR Part 280 went into effect in December 1988. These rules have forced the states to begin aggressive inventories of the leaking pipes, connections, and tanks that were silently polluting underground water sources. Fire inspectors, state environmental inspectors, insurance company risk-management advisers, and professional associations all passed the word that underground tank removal and replacement would be necessary. Knowledge of the issue, if not of the detailed methods for tank removal and remediation, is widespread among managers of environmental affairs.

In December 1988, federal EPA rules for these hazardous-liquids underground tanks split the regulatory status of tanks into new and old tanks. The previously installed tanks would need leak-detection systems and, where appropriate, removal of the tank and cleanup of underlying soil. New tanks were required to meet stringent design and installation norms so that there would be assurance that newer tanks would give adequate indications of future problems, before an environmental-adverse effect from a leak could result. Emphasis in the new design requirements is on corrosion protection and the avoidance of tank spills.

Specific duties of the environmental manager include:

1. inventorying, or maintaining records of an existing inventory, for each underground tank
2. knowing what is or was stored in the tank, even if it has been abandoned and filled with soil
3. overseeing inventory control on the petroleum, hazardous materials, or other regulated fluids kept in tanks, or maintaining a modern measurement system such as passive-acoustic array devices or automatic tank gauging
4. reporting leaks promptly to state officials as required under state underground tank rules
5. planning and implementing a safe removal operation for an older tank that no longer meets the requirements
6. planning and overseeing installation of new tanks with operating equipment capable of detecting leaks

One reason why tanks can be a hot topic is that corrective plans for leaky tank removal must be shared with neighbors via public-comment processes. Permanent closure requires a groundwater sampling effort to determine where and how much of a leak occurred. A public hearing on removal plans is a forum for critics who charge neglect and sloppy management had caused the leaks to contaminate neighboring properties, wells, and aquifers. Most institutions will have an experienced consultant act as the planner and public presenter for the community hearing, in the hope that lessons learned in other publicly challenging scenarios will make the engineering consultant more skilled at dealing with these issues. And after an old tank is closed, rules require retention of records for three years; but a prudent institution, wary of future cleanup demands for that site's later uses, will hold the basic files of the tank investigation for an indefinite time.

States have adopted their own implementing rules as part of the state assumption of responsibility for RCRA programs, including the underground-storage-tank program. To determine the terms of the underground-storage-tank rules that apply in your state, ask the state-level association that represents your interests in your state capital, such as a hospital council, to obtain and then disseminate the current operating rules of the state's fire marshal or environmental-management agency regarding tanks.

The manager responsible for environmental compliance, on a limited budget, might be tempted to use available workers to pull out the old tanks or bury those that are no longer in use. This is not wise; RCRA penalties are severe and violations can bring individual criminal charges (Wiener and Bell 1994). The institution should obtain the services of an experienced contractor who has site-assessment equipment, familiarity with groundwater-leakage remedial engineering, and especially, tank-removal expertise.

Fear of making things worse should not deter the manager from acting; rather, a cautious and experienced contractor to "do it right" is the best response to worries about tank removals. The task of removal has a propensity for spillage of residues of liquids, and this can worsen ground contamination if it is not done properly. Measurement and documentation that can meet the RCRA standards are best handled by an experienced removals contractor.

The institution's insurance managers should discuss with the environmental manager how the million-dollar state-required financial responsibility assurances can be established. Though state and federal institutions are exempt from the requirement to show financial responsibility, all others, including nonprofit owners or operators, are responsible. In the case of an operator of the tank separate from the owner of the tank, for example, a leased bus-refueling operation on land owned by a college, financial responsibility is very important because each party, owner, and operator is jointly and severally liable.

Because most institutions will have some form of environmental-agency inspection at least annually, underground-tank questions are likely to be asked.

The managers who are trained to conduct tours and respond to questions from inspectors should be told what to expect and where to find the relevant files of data. Under court decisions interpreting the RCRA law, decided by federal courts, inspections can be performed without a warrant (*V-1 Oil* v *Wyoming Dept. of Env. Quality* 1990).

Because property transfers usually involve certifications concerning the current or past status of underground storage tanks (ASTM 1993), the prudent manager who works with acquired or donated land should be certain to assure that competent surveys were done and that the presence of leaks or contaminant spills were reported in compliance with state tank-removal regulations. It is not prudent to accept land into the institution's ownership without knowing its storage-tank conditions, among other things. Insurance carrier inspections will also have the tank-removal question on their list of audit questions, so anticipating the tank issue will help the safety manager be prepared for this very likely question.

Finally, the penalties for the institution and the individual compliance manager must be considered. Penalties for mishandling one's storage-tank operations can be quite severe. Failing to provide a required notice may result in $10,000 per-day civil penalties, for each tank on which no notice was provided as required (or on which false reports were made, in which case criminal charges are extremely likely as well as the civil penalty). Citizen suits by angry neighbors against the institution for violations of the RCRA law are another problem that must be evaluated and, if possible, avoided (Greve and Smith 1992).

12.4   WATER—SAFE DRINKING WATER ACT RULES

The 1974 federal Safe Drinking Water Act (SDWA) applies to the institution if it maintains a "public" drinking-water system, for example, a college supplying water from its own wells, rather than merely connecting to a privately owned or municipal water system. The SDWA is often cited as a stringent regulatory control law imposing demands that are as tight as feasible to reduce exposures to potentially harmful contaminants (Percival et al. 1992, 437).

The basic question for the institutional manager to ask is "are we covered by this law as a 'public' water source"? If the system serves at least 15 service connections and at least 25 individuals, it will likely be covered. If the answer is no, then the system is not covered, and the institution escapes SDWA obligations for testing. If it is covered as a public water system then the institution must comply with national standard levels of contaminants, have a testing program for contaminants, maintain records, notify customers of discoveries of contamination in the water, and avoid any exceedance of SDWA maximum contaminant levels, for example, for lead.

The SDWA's requirements are extremely detailed and quite limiting to the entity that becomes a public water supply, so this text cannot do full justice to the particular demands of that law. For the institutional manager, drinking-water regulation should be part of the overall contamination-avoidance mission of the safety-engineering team. It is an engineering challenge to actually improve the quality of incoming municipal water and to deliver it safely to the drinking-water user (Stoner 1982, 88).

12.5 HAZARDOUS RELEASES—TOXICS RELEASE INVENTORY

Hospitals and educational institutions are not subject to mandatory reporting of chemical releases under the EPA's Toxic Release Inventory (TRI) program. That well-publicized annual reporting program makes public the type and amount of chemicals that a regulated facility emits, releases, or transfers to off-site disposals. The educational and hospital entities were omitted because these institutions are classified for government statistical-reporting purposes outside of the set of manufacturing entities that Congress targeted when it wrote the reporting laws.

In the event of a spill of an "extremely hazardous" chemical listed by EPA for emergency-release reporting, above the quantity at which a report is required, immediate telephone notice to the local emergency center (usually a fire department) and to the National Response Center is required. An example would be a fire that causes an explosion in an agricultural college's ammonia tanks. Notifications should be made immediately, whether or not the eventual legal review finds that one notification was not legally required. When in doubt—and faced with a release of a harmful gas, smoke, or liquid—the environmental professional is always more protected by a decision to report the release to the local emergency authorities and to the National Response Center.

The environmental manager for the institution should be very interested in local TRI information for planning purposes. Nearby factories disclose their set of extremely hazardous chemicals to the local emergency-planning committee, and their toxic-release inventory to federal and state environmental agencies. Safety plans should include response scenarios if a major ammonia release or fire occurred at one of these neighboring sites. The speed with which the hospital or dormitory or classroom building could be evacuated is probably known from fire drills, but in some chemical-emergency situations, the best response to a cloud of gas will be to have an emergency shutdown of ventilation intakes and a rapid closing of doors and windows. Using the available data about nearby sites can make the emergency-response plan of the institution more realistic, taking into account the external threats from larger-scale industrial accidents in the local area.

12.6   RADIATION—HANDLING AND DISPOSAL
OF RADIOACTIVE ITEMS

Unpleasant surprises can ruin the manager's day, and radioactive materials are a source of many surprises and ruined days (Reinhardt and Gordon 1991; Stoner 1982, 128; Shapiro 1981, 161–162). Most hospital and university safety managers find the disposal of low-level radioactive waste and contaminated materials to be an especially daunting task. Inspections can be very demanding for an aspect of the safety manager's job that is not often a top priority. The handling of low-level radioactive waste is an intensely controlled and regulated task, subject to tight inspection requirements and significant civil penalties. Prudent hospitals approach this topic with careful regard for occupational exposure, waste-container marking, and environmentally sound disposal (Reinhardt and Gordon 1991, 21). The cost of radioactive-waste disposal is high, but the alternative of ignoring the regulatory issue and just disposing of spent radionuclides carelessly is not an option, since bad publicity linking careless college administrators and radiation leaks could be devastating, and criminal penalties can be applied to knowing mishandling of radioactive materials.

The federal Nuclear Regulatory Commission (NRC) has exclusive control over radioactive materials and their wastes, under the Atomic Energy Act (AEA). That statute specifically covers medical uses of "special nuclear materials." Federal EPA has some concurrent power over mixtures of radioactive and hazardous wastes (Reicher 1993) but the jurisdictional lines are not clear (Reinhardt and Gordon 1991, 21). States have workplace safety roles, but the specific control of radioactive material is an exclusive NRC matter which the other governmental entities cannot administer or regulate. A college might pass four EPA inspections and six OSHA inspections but then be fined heavily for radioactive equipment safety violations found by an NRC inspector in a physics lab.

The institution that intends to have radioactive materials for use in research or patient care must create a radiation-safety program. Then it should designate at least one of its science-trained managers to be the radiation safety officer who will attend classes, obtain necessary educational materials, and account for the radioactive materials.

Licenses from the NRC may be obtained in a number of ways. By NRC rules, a "general license" applies to any hospital or physician to obtain and use the least serious of "by-product materials." A similar general license permits pharmaceutical makers to make radiopharmaceuticals and reagent kits using radioactive materials. For the hospital, very detailed rules limit the extent to which radionuclide materials can be used or distributed.

Environmental control of wastewater discharges containing low-level radioactive materials, such as rinse water from materials used in radiological procedures, is managed by the federal NRC. A state or locality cannot impose a more stringent standard as to radiation, though the local or state officials may regulate "any other toxic or hazardous properties" of the low-level radioactive wastes. The localities also can regulate nonsafety issues such as economic controls upon NRC-licensed facilities. Shipment of nuclear materials is closely regulated by the NRC. The occupational safety requirements for monitoring of radiation exposures are also a heavily federal issue.

12.7 NOISE—BUILDING, STUDENT, AND WORKPLACE NOISE

Noise is a liability and a regulatory issue for institutions such as hospitals and universities (Stoner 1982, 82). The federal Noise Control Act is available, and has some limiting effects on local laws affecting noise controls (*Burbank* v *Lockheed Air Terminal* 1973; *Southern Pacific* v *EPA* 1993), but the federal Act has little benefit for the institution's safety planners. The Act's power to set standards for noise in construction equipment, engines, electrical equipment, and so on, has gone largely unused by federal EPA over the years since its adoption in 1972 (Shapiro 1992).

Limited preemption of local laws is included in the federal Noise Control Act (*Southern Pacific* v *EPA* 1993). This means that local laws cannot exist where the federal law imposes a control. But this only covers such issues as aircraft engine regulations, which the federal Act specifically covered. Like other political accommodations with neighbors, a facility's voluntary noise-control efforts can help reduce friction with community residents and elected officials representing constituents around the facility.

Objections to college and university expansion in a residential neighborhood will usually include fears of worsened noise conditions as a factor in zoning objections (*Glenbrook Road Assn.* v *DC Board of Zoning* 1992). College students keep unusual hours and have unusual late-night amusement patterns. For hospitals, ambulance and delivery truck noise is a frequent objection. Noise can be a reason for objections in hospital locations (*Clinton Community Hospital* v *So. Maryland Medical Center* 1975). For schools, football games and boisterous student outdoor activities tend to be a source of local friction.

The manager of the safety function also has to deal with OSHA and workers' compensation problems related to noise. Hearing loss is an expensive compensation problem, and because the research or educational institution is not like an assembly-line factory, the individual department or unit worker's noise problems are not simply controlled by one factory-wide version of a hearing-conservation program. The standard industrial-hygiene texts describe the baseline audiogram and other monitoring and corrective measures. It is sufficient here to note that OSHA citations for excessive workplace noise may be issued when noise levels remain above acceptable norms for a time-weighted period.

12.8 MEDICAL EXPERIMENTATION

The safety aspects of medical experimentation at a hospital or university are usually well managed by clinical research professionals, with the help of a product-regulatory manager from the cooperating company that is interested in the product being tested. An institutional review board typically acts as the gatekeeper to assure that patients' rights are protected (Note Institutional Review Boards 1988). Much of the interaction with the Food and Drug Administration or other government agency is entrusted to the company whose drug, device, pesticide, and so on is being researched by experts at the hospital or educational institution. Because other sources already cover the product-related regulatory issues very extensively (O'Reilly 1993, Chap. 23), this chapter will devote its coverage to the role of the institution's safety manager.

The simplest approach for the safety manager should be to discuss, before research begins, the waste that will be produced from the upcoming series of tests, and how the waste will be secured and characterized (needle "sharps," bloody drapes, human tissue, flammable liquids) so that janitorial staff and nearby workers are protected from accidents (Reinhardt and Gordon 1991). Sewer-permit conditions imposed on the institution, which limit certain discharges, should be explained to the researchers if a part of the research plan involves vessel cleanout, process-water disposal, or other sewering of part of the experimental material.

When an experimental drug or material that is truly new is being introduced for the first time, the environmental manager should consider whether a laboratory stack or fume hood needs a modified air-emission permit and/or whether a sewer permit should be modified for water-based test articles that are beyond terms of the wastewater-treatment permit.

The more serious risks from handling or being exposed to the test substance, for example, workers spraying an agricultural research station's test crop with a pesticide, increase the institution's risk for civil-damage lawsuits if the work is not done with appropriate attention to details (*Ferrebee* v *Chevron Chemical* 1984).

Federal or state OSHA inspectors examining the safety of the programs at the institution will regard the experimental nature of the new chemical as no excuse for failing to perform safety training. They will likely focus on safety training of laboratory workers under 29 CFR § 1910.1450, and on informed-consent paperwork covering the patients who were exposed to the experimental material, as required under 21 CFR § 50.20 (Troyer and Salman 1986; O'Reilly 1993). Managers need to consider the right amount of safety training about risk-avoidance should be part of the institution's hazard-communication program. Workers, including students, may be affected by exposure to the experimental chemical and this concern should be met by adequate training. It is significant that if the new innovative material in the experiment involves a non-FDA chemical, it is likely that there must be worker notice and training under the EPA's Toxic Substances Control Act rules concerning the "research and development exemption from premanufacture notification."

The safety manager should be aware that fraud in medical research often involves the concealment of adverse effects data in reports of the experimentation (O'Reilly 1990, 393). Signatures and patient identifiers are part of the institution's defense against injury lawsuits. Accurate documentation of the results of research and honest compilation of patient medical records should be universally respected, but where they are not followed, the safety manager has a personal stake in preserving the credibility of the hospital or university. Publicity about fraud in one area invites regulators to snoop out records from other areas in search of a pattern of deception. Such a cycle of critical inspections looking for "something bad" is destructive and hard to stop once it begins. Though the safety manager is not tasked to police the accuracy of experimentation, falsity in research colors the credibility of the institution to its community and to other regulatory bodies.

REFERENCES

ASTM Inc. 1993. Environmental Site Assessment, Standard 1527-93 Philadelphia: ASTM Inc.

Burbank v Lockheed Air Terminal Inc., 411 U.S. 624 (1973).

Clinton Community Hospital Corp. v Southern Maryland Medical Center, 510 F.2d 1037 (4th Cir., 1975), cert. den. 422 U.S. 1048 (1975).

Ferrebee v Chevron Chemical Co., 736 F.2d 1529 (D.C. Cir., 1984), cert. den. 469 U.S. 1062 (1984).

Glenbrook Road Assn. v District of Columbia Board of Zoning Adjustment, 605 A.2d 22 (D.C. Ct. App. 1992).

Greve, Michael, and F. Smith. 1992. Environmental Politics: Public Costs, Private Rewards.

In Re Villanova University, 1988 TSCA LEXIS 6, EPA Dkt. TSCA-III-159 (1988).

Note (Student Work). 1988. Institutional Review Boards in the University Setting. Journal of College and University Law 15: 185.

O'Reilly, James. 1990. More Gold and More Fleece: Improving the Sanctions Against Medical Research Fraud. *Administrative Law Review* 42: 393.

O'Reilly, James. 1993. *Food and Drug Administration.* 2d ed. Colorado Springs: McGraw-Hill/Shepards.

O'Reilly, James, Kyle Kane, and Mark Norman. 1993. *RCRA & Superfund Practice Guide.* 2d ed. Colorado Springs: McGraw-Hill/Shepards.

Pacific Gas & Electric Co. v *California Energy Commission,* 461 U.S. 190 (1983).

Percival, Robert, Alan Miller, Christopher Schroeder, and James Leape. 1992. *Environmental Regulation: Law, Science and Policy.* Boston: Little Brown.

Reicher, Dan. 1993. Nuclear Energy and Weapons. In *Sustainable Environmental Law,* ed. Celia Campbell-Mohn, Barry Breen, J. William Futrell, p. 954. St. Paul MN: West Publishing.

Reinhardt, Peter, and Judith Gordon. 1991. *Infectious and Medical Waste Management.* Boca Raton, FL: Lewis.

Shapiro, Fred. 1981. *Radwaste: A Reporter's Investigation of a Growing Nuclear Menace.* New York: Random House.

Shapiro, Sidney. 1992. Lessons from a Public Policy Failure: EPA and Noise Abatement. *Ecology Law Qtly* 19: 1.

Southern Pacific Railroad C. v *EPA,* 9 F.3d 807 (9th Cir., 1993).

Stoner, David. 1982. *Engineering a Safe Hospital.* New York: John Wiley & Sons.

Troyer, Glenn, and Steven Salman. 1986. *Handbook of Health Care Risk Management.* Washington DC: Aspen.

United States Code, Title 42, various sections.

U.S. EPA. 1994. 40 Code of Federal Regulations 280–281.

U.S. Food & Drug Administration. 1994. 21 Code of Federal Regulations 50.20.

U.S. Nuclear Regulatory Commission. 1994. 10 Code of Federal Regulations Parts 20, 33.

U.S. Occupational Safety & Health Administration. 1994. 29 Code of Federal Regulations 1910.95, 1910.1200, 1910.1450.

University of Pittsburgh. 1980. *Occupational Safety & Health Decisions (CCH)* para. 24,240 (Rev. Com. 1980).

V-1 Oil Co. v *Wyoming Dept. of Environmental Quality,* 902 F.2d 1482 (10th Cir. 1990).

Wiener, Robin, and Christopher Bell. 1994. *RCRA Compliance and Enforcement Manual.* 2d ed. Colorado Springs: Shepards/McGraw-Hill.

Chapter 13

Art and Theater Department Hazards

13.1 BACKGROUND

Institutions that frequently deal with children in a supervisory capacity (e.g., primary and secondary school systems, day-care programs, or health-care programs) often make use of art or theater programs as part of their curriculum. Some of the art materials commonly used in such programs can pose significant health risks; facility owners and operators should be aware of the potential risks.

Though their students may not require the same level of supervision, college or university art and theater departments tend to employ a much wider variety of materials and techniques, creating a much broader range of potential health risks. The background or training of art instructors may not include detailed knowledge of art-related health hazards.

Similarly, many arts-and-crafts courses taught in health and continuing-education or rehabilitation/therapy settings are run by volunteers who have not been made aware of art-materials health hazards in any systematic manner. Their students—typically children and ill or post-operative patients—can be especially vulnerable to the kinds of health hazards that some art materials can pose.

Because most of the risks found in theater department programs are a subset of those encountered by art departments, this section focuses on what are traditionally thought of as "art" materials and activities, though many of them are common in theater contexts as well.

13.2 EXPOSURE

Toxic substances can enter the human body in three different ways: skin absorption, inhalation, and ingestion.

Skin Absorption

Many art materials, such as acids, solvents, dyes, or dusts can damage skin tissue with even a minimal amount of contact. Some substances can also penetrate the skin and enter the bloodstream directly; repeated exposures can lead to serious kidney, liver, and heart damage.

Inhalation

Gases, dust particles, and aerosol mists can damage the entire respiratory tract, penetrating even to the deepest regions of the lungs. Once in the lungs, many substances pass on into the bloodstream, where they are carried to tissue systems throughout the body. Effects range from acute, immediate harm (headaches, dizziness, nausea) to low-level chronic tissue irritation to, in some cases, permanent damage.

Ingestion

Art materials are usually ingested accidentally—food, beverages, cigarettes, cups, paintbrush handles—all can deliver hazardous materials to the mouth. Airborne dusts and mists, spilled liquids, and friable solids can also make their way onto plates, glasses, rags, and any other item left in the work area.

13.3 SUSCEPTIBILITY

Susceptibility to toxins in hazardous art materials will vary with the physical characteristics of the students and artists. When compared to most adults, children under the age of 12 are especially vulnerable to toxic materials. Because of their smaller body size and mass, even a small amount of toxin can have a very large effect. Their metabolism is generally faster, which causes their bodies to absorb a given toxin more quickly. Because their immune systems are still developing, they are less able to process unusual toxins and are more vulnerable to developmental damage. Finally, children are more vulnerable than adults simply because they are less likely to understand the risks and the consequences of exposure. Adults who are pregnant, allergy sufferers, heavy smokers, or drug users can also be particularly vulnerable.

13.4   EVALUATING YOUR ACTIVITIES

In evaluating the health risks of different aspects of your operations, a good place to start is with the material-safety-data-sheet (MSDS) system (EPA, § 1910.1200). An MSDS contains information about a substance's chemical formula, potential health hazards, and directions for its proper use. Occupational Safety and Health Administration (OSHA) regulations require manufacturers and importers of chemical substances to include MSDSs for hazardous substances. Operators of art/theater programs should request these sheets from their art/theater materials distributors, who are under an affirmative duty to pass along this information.

Instructors and, depending on their financial arrangements, some students, may qualify as employees, as discussed in Chapter 2. Their workplaces are thus governed by OSHA regulations. Owners and operators of these workplaces need to take steps to ensure that both their employees' handling of potentially hazardous art materials and the management of these programs by their supervisors conform to these regulations. OSHA's Hazard Communication program sets forth general health and safety guidelines for all operations, including art/theater workplaces. Local OSHA branch offices can assist facility owners and operators in assessing their obligations.

Finally, some art materials may contain substances that are considered "hazardous" under the Resource Conservation and Recovery Act (RCRA) (EPA, §§ 261.20 –.24, .30 –.33). So long as an operation generates less than 220 pounds (100 kilograms) of these hazardous substances in one month, it is considered to be "conditionally exempt" from RCRA's disposal requirements, discussed in Chapter 8, and may simply dispose of the waste at a recycling center or a state-permitted solid-waste landfill (EPA, § 261.5). If, however, an operation generates more than 220 pounds (100 kilograms) of these wastes in a month, it must obtain an EPA identification number (EPA, § 262.12) from its state environmental agency and dispose of the waste according to RCRA's hazardous-waste-disposal regulations (EPA, §§ 268.40 and 268.45). This includes treatment of the waste and specific disposal procedures. Please see Sections 10.1–10.2 on solid-waste disposal for a more detailed discussion of these requirements.

13.5   MATERIALS FOR CHILDREN

In 1988, Congress enacted the Labeling of Hazardous Art Materials Act (LHAMA) (USC, § 1277). LHAMA's primary purpose is to encourage the safe use of art and craft materials by students and artists "of any age." It also codified the American Society for Testing and Materials (ASTM) D-4326 document as the Consumer Product Safety Commission (CPSC)'s "chronic health hazard" art materials labeling standard. The LHAMA standards require producers/repackagers of art materials to submit their products' compositional formulas to toxicologists for a safety review. If the toxicologists determine that the product presents a potential chronic health hazard, they will specify which product labels the producers/repackagers must use. All art materials requiring chronic-hazard labeling must bear appropriate warning labels, such as "May be harmful if swallowed" or "Keep out of reach of children" (CFR, § 1500.14(b)(8)(i) and USC, §1277(b) and (d)).

The CPSC is authorized to enjoin the purchase of any art material requiring LHAMA labeling which is for use by children in pre-kindergarten, kindergarten, or grades 1 through 6. The CPSC focuses its enforcement efforts on those art materials (i.e., paint, glue, glaze, ink) that are most commonly used and, therefore, present the greatest risk of public harm. It is does not take enforcement actions against products that are not normally marketed for art purposes or that are "tools of the trade"—that is, canvases, paintbrushes, chisels, paper, lathes, and so on.

13.6   ART MATERIALS SEGMENTS

Brief descriptions of the primary health hazards associated with the major art segments follow. These are designed to serve as a guide or a prompt in assessing a facility's responsibilities in this area.

Photography

Photographers are exposed to a wide variety of hazardous chemicals and must commonly work in poorly ventilated areas. Developers, stop baths, and fixing baths commonly must be prepared prior to use, creating a second opportunity for exposure to these chemicals, often in their raw state. Representative activities include color and black-and-white photography, including negative development, printmaking, and print enhancement.

Glass/Ceramics/Pottery

The health risks associated with these activities include inhalation of silicaceous dusts and mists, the application of glazes and colorings during processing, and the final firing and/or curing of these pieces with intense heat. Glassworking, in particular, can involve the use of lead compounds, other toxic colorants (i.e., cadmium, nickel, chromium) and strong acids. Representative activities include mixing glazes and colorants, their application to pottery and ceramics, and the firing of these pieces as well as glassblowing and stained-glass manufacture.

Painting/Drawing/Printmaking

Many of the materials used in these areas are either flammable, corrosive, or contain potentially harmful hydrocarbon and/or metal constituents. Representative activities include intaglio and lithographic printing; acid and photo etching; silk-screening; oil, acrylic, airbrush and spray painting; varnishing and lacquering.

Textiles

The largest health concerns in the textiles area are fiber dusts and toxic dyes. Inhalation of fibers can cause chronic irritation of the respiratory tract, and exposure for prolonged periods can lead to permanent damage. Toxic dyes can enter the bloodstream through any of the three body pathways, provoking varying degrees of harm. Representative activities include processing, dyeing, carding, and spinning of fabrics.

Metalworking/Jewelry

Persons taking part in these activities can be exposed to toxic fumes, chemical burns, and metal dusts, as well as infrared and ultraviolet radiation, electric shock, and fire. Representative activities include welding, casting, forging, electroplating, and finishing of metal products and jewelry.

Sculpture

Sculpture can involve the use of materials from a wide range of substances: metal, wood, stone, clay, and plastics. Though each of these substances carries its own particular set of health hazards, they all focus primarily on respiratory exposure to dusts and mists or larger fragments and splinters of material. Representative activities include sculpting metal, wood, stone, clay, and plastics as well as the production of multimedia and multiprocess pieces.

13.7 PREVENTATIVE STRATEGIES

Three principles of art-materials safety are common to all of the art activities discussed above:

    a. *Substitution*—Whenever possible, use less-toxic or nontoxic materials, e.g., use mineral spirits instead of turpentine or lacquer thinner, use lead-free rather than leaded pottery glazes.

    b. *Safety Equipment*—Always require the use of appropriate safety equipment, e.g., gloves, eye goggles, respirators, and ensure that art materials are stored properly.

    c. *Procedures*—Implement practices that minimize the materials' health hazard, e.g., use wet clays instead of dry in order to cut down on ambient dust, hand-brush paints rather than spray them on.

As you analyze your facility's use of materials, use these three principles to guide you in developing your safety guidance policies.

13.8 SPECIFIC PRECAUTIONS

Good practices for art- and theater-department hazards mimic the engineering, personal protective equipment, and work-practices standards associated with a workplace analysis under general industrial hygiene criteria. In each instance, the task is to implement practices appropriate for the type of hazard and the maturity and training of the involved students.

Respiratory Protection

Adequate ventilation is essential. This can range from normal ventilation to open windows and room fans to exhaust hoods or spray booths. In heavily contaminated environments, local exhaust ventilation systems may be required.

When handling powdered substances, or engaging in airbrushing or spray-painting, respiratory protection ranging from face masks to respirators may be warranted. Respirators require care in selection, fitting, and maintenance, and may not be medically sound for persons with respiratory ailments.

Other Personal Protective Equipment

Depending on the hazards involved, gloves, goggles, face shields, hearing protectors, and even fire-resistant clothing and infrared goggles may be required to guard against respiratory, skin, and eye hazards. Eye protection, gloves, and protective clothing should always be worn when working with strong acids or bases. Eyewash fountains and showers should be readily accessible.

Work Practices and Building Maintenance

Acids, pigments, colorants, chemical solutions, and other powdered or liquid materials should remain covered when not in immediate use. Smoking should not be permitted in these work environments.

Adequate cleaning should prevent dust and material buildup. Depending on the activities and materials involved, surfaces might be damp-wiped or damp-mopped, or vacuumed, perhaps with a High-Efficiency-Particle-Absorbing (HEPA)-type filter vacuum to avoid simply recirculating dust.

Finally, materials must be disposed of properly. The nature of the material will determine whether it is an ordinary solid waste, such as clay, or a hazardous waste that must be handled accordingly. The chapters on worker safety and hazardous-materials disposal provide more detail on these issues.

REFERENCES

16 Code of Federal Regulations § 1500.14(b)(8)(i)(E).
29 Code of Federal Regulations § 1910.1200.
40 Code of Federal Regulations § 261.5.
40 Code of Federal Regulations § 261.20–.24.
40 Code of Federal Regulations § 261.30–.33.
40 Code of Federal Regulations § 262.12.
40 Code of Federal Regulations § 268.40.
40 Code of Federal Regulations § 268.45.
15 United States Code § 1277(b).
15 United States Code § 1277(d).

Chapter *14*

The Safe Drinking Water Act

14.1 APPLICABILITY TO EDUCATIONAL
AND HEALTH-CARE FACILITIES

Drinking water in the United States is primarily regulated under the Safe Drinking Water Act (SDWA) (USC, § 300(f) et seq.), which is administered by the U.S. Environmental Protection Agency (EPA). The SDWA is predominantly a state-administered program, although state drinking-water requirements generally must be no less stringent than federal requirements (CFR 142 Part B). Under the SDWA, EPA is required to develop regulations establishing water-quality standards that must be met by all Public Water Systems (PWSs). A school or hospital is a PWS if it provides water to 25 people for 60 days per year. However, facilities or buildings that merely distribute water from their local water district generally are not PWSs (USC, § 300g).

In addition to providing the foundation for regulation of PWSs, the SDWA also serves as the basis for other regulatory programs including EPA's Underground Injection Control (UIC) Program. The UIC is a permitting program that regulates any subsurface placement of fluids by well injection that might endanger drinking water sources. EPA has established regulations outlining effective state UIC programs; if they so choose, states may assume primary enforcement responsibility for these programs.

The SDWA also grants EPA the authority to implement other groundwater protection programs, including the Wellhead Protection Program (WPP) and the Critical Aquifer Protection Area Demonstration (CAPAD) program.

Drinking-water requirements are not often a part of the institutional manager's personal experience. EPA's SDWA Hotline (800-426-4791) is an excellent resource for general information relating to the SDWA. Questions on home water-filtration systems can be directed to the National Sanitation Foundation (NSF) at (313) 769-8010. Questions concerning bottled water should be directed to the Food and Drug Administration (FDA) at (301) 443-4166. Additional background material on the SDWA is available from the Association of State Drinking Water Administrators (ASDWA) at (202) 293-7655.

14.2   REGULATION OF WATER SYSTEMS

A Public Water System (PWS) is any system that provides piped "water for human consumption" and which has at least 15 service connections, or regularly serves an average of at least 25 people daily at least 60 days per year. (EPA, § 141.2) "Water for human consumption" includes water that is used for bathing and showering, cooking, dishwashing, and maintaining oral hygiene. Despite use of the term "public," a water system need not be publicly owned to be considered a PWS.

EPA divides PWSs into community water systems (CWS), nontransient noncommunity water systems (NTNCWS), and transient noncommunity water systems (TNCWS). A CWS is a water system that serves at least 15 service connections used by year-round residents, or regularly serves at least 25 year-round residents. NTNCWSs regularly serve at least 25 of the same people over six months of the year. Colleges, hospitals, schools, and factories that serve water to 25 or more of the same people for six or more months of the year are examples of NTNCWSs. TNCWSs are noncommunity water systems that do not regularly serve at least 25 of the same persons over six months per year. TNCWSs often include, for example, restaurants, gas stations, campgrounds, and churches.

As discussed in greater detail below, National Primary Drinking Water Regulations (NPDWRs), the principal regulatory provisions of the SDWA, generally apply to PWSs. However, certain PWSs are not subject to NPDWRs. Specifically, a PWS is not subject to NPDWRs if it (EPA, § 141.3):

1. consists only of distribution and storage facilities and does not have any collection and treatment facilities
2. obtains all of its water from, but is not owned or operated by a PWS subject to NPDWRs
3. does not sell water to any person
4. is not a carrier that conveys passengers in interstate commerce

This chapter relates to tap water only. Bottled water is a "food" regulated separately by the Food and Drug Administration (FDA 103.35).

14.3   DRINKING-WATER REGULATIONS

Primary Standards/MCLs

National Primary Drinking-Water Regulations (NPDWRs) (see Table 14–1) apply to all PWSs (except as noted above) and establish the minimum acceptable standards for public drinking water in the United States (USC, § 300g-1). These standards are realized through the establishment of maximum contaminant levels (MCLs) and required treatment techniques. Specifically, the SDWA requires EPA to publish "maximum contaminant level goals" (MCLGs) for contaminants that, in EPA's judgment, may have an adverse effect on public health and that are known or anticipated to occur in PWSs. MCLGs are to be set at a level at which "no known or anticipated adverse effects on the health of persons occur and which allows an adequate margin of safety" (USC § 300g-1(b)(4)). When EPA publishes an MCLG, which is a nonenforceable health-based goal, EPA must also promulgate a NPDWR that includes either:

1. a maximum contaminant level (MCL), which must be set as close to the MCLG as "feasible," or
2. a required treatment technique (USC, § 300g-1(b))

Each NPDWR must describe the best available technology (BAT), treatment techniques, and other feasible means for meeting its MCL. NPDWRs also include monitoring, analytical, and quality-assurance requirements.

An MCL is the maximum level of a contaminant (typically expressed in milligrams per liter (mg/L)) permitted in water delivered to a user by a PWS. When determining the MCLG for a contaminant, EPA looks to established reference doses (RfD) for toxic substances. RfDs are human exposure levels at which no observable "adverse effects" are expected. Once it has established an MCLG, the Agency must then consider the best available technology (BAT) (e.g., air stripping, sand filtration, ultraviolet disinfection) for treatment of that contaminant. The contaminant concentration remaining in drinking water after BAT treatment, as well as the "practical quantification level" (PQL) (the lowest concentration that most testing laboratories can consistently detect), largely determine the contaminant's MCL. Thus far, EPA has established MCLs for almost 100 contaminants.

Every three years, EPA must publish a "priority list" of at least 25 contaminants that are "known or anticipated to occur" in PWSs and which, due to their hazardous or toxic characteristics, "may require action" on the part of the PWS operator (USC, § 300g-1(b)). The list includes contaminants such as pesticides, solvents, cleaners, and petroleum products. These additions to the Drinking

Table 14-1. Table of National Primary Drinking-Water Standards

Contaminants	MCLG (mg/L)	MCL (mg/L)	Potential Health Effects from Ingestion of Water	Sources of Contaminant in Drinking Water
Fluoride	4.0	4.0	Skeletal and dental fluorosis	Natural deposits; fertilizer, aluminum industries; water additive
Volatile Organics				
Benzene	zero	0.005	Cancer	Some foods; gas, drugs, pesticide, paint, plastic industries
Carbon Tetrachloride	zero	0.005	Cancer	Solvents and their degradation products
p-Dichlorobenzene	0.075	0.075	Cancer	Room and water deodorants, and "moth-balls"
1,2-Dichloroethane	zero	0.005	Cancer	Leaded gas, fumigants, paints
1,1-Dichloroethylene	0.007	0.007	Cancer, liver and kidney effects	Plastics, dyes, perfumes, paints
Trichloroethylene	zero	0.005	Cancer	Textiles, adhesives and metal degreasers
1,1,1-Trichloroethane	0.2	0.2	Liver, nervous system effects	Adhesives, aerosols, textiles, paints, inks, metal degreasers
Vinyl Chloride	zero	0.002	Cancer	May leach from PVC pipe; formed by solvent breakdown
Coliform and Surface Water Treatment				
Giardia lamblia	zero	TT	Gastroenteric disease	Human and animal fecal waste
Legionella	zero	TT	Legionnaire's disease	Indigenous to natural waters; can grow in water-heating systems

Standard Plate Count	N/A	TT	Indicates water quality, effectiveness of treatment	
Total Coliform*	zero	<5%+	Indicates gastroenteric pathogens	Human and animal fecal waste
Turbidity*	N/A	TT	Interferes with disinfection, filtration	Soil runoff
Viruses	zero	TT	Gastroenteric disease	Human and animal fecal waste
Phase II-Inorganics				
Asbestos (>10um)	7MFL	7MFL	Cancer	Natural deposits; asbestos cement in water systems
Barium*	2	2	Circulatory system effects	Natural deposits; pigments; epoxy sealants, spent coal
Cadmium*	0.005	0.005	Kidney effects	Galvanized pipe corrosion; natural deposits; batteries, paints
Chromium* (total)	0.1	0.1	Liver, kidney, circulatory disorders	Natural deposits; mining, electroplating, pigments
Mercury* (inorganic)	0.002	0.002	Kidney, nervous system disorders	Crop runoff; natural deposits; batteries, electrical switches
Nitrate*	10	10	Methemoglobulinemia	Animal waste, fertilizer, natural deposits, septic tanks, sewage
Nitrite	1	1	Methemoglobulinemia	Same as nitrate; rapidly converted to nitrate
Selenium*	0.05	0.05	Liver damage	Natural deposits; mining, smelting, coal/oil combustion

(continued)

Table 14-1. Table of National Primary Drinking-Water Standards (continued)

Contaminants	MCLG (mg/L)	MCL (mg/L)	Potential Health Effects from Ingestion of Water	Sources of Contaminant in Drinking Water
Phase II - Organics				
Acrylamide	zero	TT	Cancer, nervous system effects	Polymers used in sewage/wastewater treatment
Alachlor	zero	0.002	Cancer	Runoff from herbicide on corn, soy-beans, other crops
Aldicarb*	0.001	0.003	Nervous system effects	Insecticide on cotton, potatoes, others; widely restricted
Aldicarb sulfone*	0.001	0.002	Nervous system effects	Biodegradaton of aldicarb
Aldicarb sulfoxide*	0.001	0.004	Nervous system effects	Biodegradation of aldicarb
Atrazine	0.003	0.003	Mammary gland tumors	Runoff from use as herbicide on corn and noncropland
Carbofuran	0.04	0.04	Nervous, reproductive system effects	Soil fumigant on corn and cotton; restricted in some areas
Chlordane*	zero	0.002	Cancer	Leaching from soil treatment for termites
Chlorobenzene	0.1	0.1	Nervous system and liver effects	Waste solvent from metal-degreasing processes
2,4-D*	0.07	0.07	Liver and kidney damage	Runoff from herbicide on wheat, corn, rangelands, lawns
o-Dichlorobenzene	0.6	0.6	Liver, kidney, blood cell damage	Paints, engine-cleaning compounds, dyes, chemical wastes
cis-1,2-Dichloroethylene	0.07	0.07	Liver, kidney, nervous, circulatory	Waste industrial extraction solvents

trans-1,2-Dichloroethylene	0.1	0.1	Liver, kidney, nervous, circulatory	Waste industrial extracton solvents
Dibromochloropropane	zero	0.0002	Cancer	Soil fumigant on soybeans, cotton, pineapple, orchards
1,2-Dichloropropane	zero	0.005	Liver, kidney effects; cancer	Soil fumigant; waste industrial solvents
Epichlorohydrin	zero	TT	Cancer	Water-treatment chemicals; waste epoxy resins, coatings
Ethylbenzene	0.7	0.7	Liver, kidney, nervous system	Gasoline; insecticides; chemical-manufacturing wastes
Ethylene dibromide	zero	0.00005	Cancer	Leaded gas additives; leaching of soil fumigant
Heptachlor	zero	0.0004	Cancer	Leaching of insecticide for termites, very few crops
Heptachlor epoxide	zero	0.0002	Cancer	Biodegradation of heptachlor
Lindane	0.0002	0.0002	Liver, kidney, nerve, immune, circulatory	Insecticide on cattle, lumber, gardens; restricted 1983
Methoxychlor	0.04	0.04	Growth, liver, kidney, nerve effects	Insecticide for fruits, vegetables, alfalfa, livestock, pets
Pentachlorophenol	zero	0.001	Cancer; liver and kidney effects	Wood preservatives, herbicide, cooling-tower wastes
PCBs	zero	0.0005	Cancer	Coolant oils from electrical transformers; plasticizers
Styrene	0.1	0.1	Liver, nervous-system damage	Plastics, rubber, resin, drug industries; leachate from city landfills

(*continued*)

Table 14-1. Table of National Primary Drinking-Water Standards (continued)

Contaminants	MCLG (mg/L)	MCL (mg/L)	Potential Health Effects from Ingestion of Water	Sources of Contaminant in Drinking Water
Tetrachloroethylene	zero	0.005	Cancer	Improper disposal of dry-cleaning and other solvents
Toluene	1	1	Liver, kidney, nervous, circulatory	Gasoline additive; manufacturing and solvent operations
Toxaphene	zero	0.003	Cancer	Insecticide on cattle, cotton, soybeans; canceled 1982
2,4,5-TP	0.05	0.05	Liver and kidney damage	Herbicide on crops, right-of-way, golf courses; canceled 1983
Xylenes (total)	10	10	Liver, kidney; nervous system	By-product of gasoline refining; paints, inks, detergents
Lead and Copper				
Lead*	zero	TT†	Kidney, nervous system damage	Natural/industrial deposits; plumbing, solder, brass alloy faucets
Copper	1.3	TT‡	Gastrointestinal irritation	Natural/industrial deposits; wood preservatives, plumbing
Phase V - Inorganics				
Antimony	0.006	0.006	Cancer	Fire retardants, ceramics, electronics, fireworks, solder
Beryllium	0.004	0.004	Bone, lung damage	Electrical, aerospace, defense industries

Cyanide	0.2	0.2	Thyroid, nervous-system damage	Electroplating, steel, plastics, mining, fertilizer
Nickel	0.1	0.1	Heart, liver damage	Metal alloys, electroplating, batteries, chemical production
Thallium	0.0005	0.002	Kidney, liver, brain, intestinal	Electronics, drugs, alloys, glass
Organics				
Adipate, (di(2-ethylhexyl))	0.4	0.4	Decreased body weight; liver and testes damage	Synthetic rubber, food packaging, cosmetics
Dalapon	0.2	0.2	Liver, kidney	Herbicide on orchards, beans, coffee, lawns, road/railways
Dichloromethane	zero	0.005	Cancer	Paint stripper, metal degreaser, propellant, extraction
Dinoseb	0.007	0.007	Thyroid, reproductive organ damage	Runoff of herbicide from crop and non-crop applications
Diquat	0.02	0.02	Liver, kidney, eye effects	Runoff of herbicide on land & aquatic weeds
Dioxin	zero	0.00000003	Cancer	Chemical production by-product; impurity in herbicides
Endothall	0.1	0.1	Liver, kidney, gastrointestinal	Herbicide on crops, land/aquatic weeds; rapidly degraded
Endrin	0.002	0.002	Liver, kidney, heart damage	Pesticide on insects, rodents, birds; restricted since 1980
Glyphosate	0.7	0.7	Liver, kidney damage	Herbicide on grasses, weeds, brush

(continued)

Table 14-1. Table of National Primary Drinking-Water Standards (continued)

Contaminants	MCLG (mg/L)	MCL (mg/L)	Potential Health Effects from Ingestion of Water	Sources of Contaminant in Drinking Water
Hexachlorobenzene	zero	0.001	Cancer	Pesticide production waste by-product
Hexachlorocyclopentadiene	0.05	0.05	Kidney, stomach damage	Pesticide production intermediate
Oxamyl (Vydate)	0.2	0.2	Kidney damage	Insecticide on apples, potatoes, tomatoes
PAHs (benzo(a)pyrene)	zero	0.0002	Cancer	Coal tar coatings; burning organic matter; volcanoes, fossil fuels
Phthalate, (di(2-ethylhexyl))	zero	0.006	Cancer	PVC and other plastics
Picloram	0.5	0.5	Kidney, liver damage	Herbicide on broadleaf and woody plants
Simazine	0.004	0.004	Cancer	Herbicide on grass sod, some crops, aquatic algae
1,2,4-Trichlorobenzene	0.07	0.07	Liver, kidney damage	Herbicide production; dye carrier
1,1,2-Trichloroethane	0.003	0.005	Kidney, liver, nervous system	Solvent in rubber, other organic products; chemical production wastes
Other Proposed (P) and Interim (I) Standards				
Beta/photon emitters (I) and (P)	zero	4 mrem/yr	Cancer	Decay of radionuclides in natural and man-made deposits
Alpha emitters (I) and (P)	zero	15 pCi/L	Cancer	Decay of radionuclides in natural deposits

Contaminant			Health Effect	Sources
Combined Radium 226/228 (I)	zero	5 pCi/L	Bone cancer	Natural deposits
Radium 226*(P)	zero	20 pCi/L	Bone cancer	Natural deposits
Radium 228*(P)	zero	20 pCi/L	Bone cancer	Natural deposits
Radon (P)	zero	300 pCi/L	Cancer	Decay of radionuclides in natural deposits
Uranium (P)	zero	0.02	Cancer	Natural deposits
Sulfate (P)	400/500	400/500	Diarrhea	Natural deposits
Arsenic*(I)	0.05	0.05	Skin, nervous system toxicity	Natural deposits; smelters, glass, electronics wastes; orchards
Total Trihalomethanes (I)	zero	0.10	Cancer	Drinking water chlorination by-products

NOTES: *Indicates original contaminants with interim standards which have been revised.
TT=Treatment technique requirement MFL=Million fibers per liter
†Action Level = 0.015 mg/L ‡Action Level=1.3mg/L
pCi=picocurie—a measure of radioactivity mrem=millirems—a measure of radiation absorbed by the body

Water Priority List (DWPL) serve as a guide to contaminants that are candidates for future regulation; EPA must promulgate NPDWRs for at least 25 contaminants on each new "priority list" within three years of its publication. By monitoring these lists, PWS operators can anticipate impending MCL and treatment-technique requirements.

NPDWRs contain specific disinfection and filtration requirements. Under the Surface Water Treatment Rule (SWTR), all systems using surface water or "groundwater under the direct influence of surface water" must disinfect their water supply (EPA, § 141.70 et seq.). The SWTR is designed to prevent exposure to viruses, bacteria, and other pathogens (Federal Register, July 29, 1994). In addition to its disinfection requirements, the SWTR also requires all affected systems to filter their water, unless they can demonstrate that they have an effective watershed protection program and can meet other specified requirements.

In 1992, EPA issued a draft "Ground Water Disinfection Rule" containing monitoring and disinfection requirements for PWSs that use groundwater that is *not* under the direct influence of surface water (Federal Register, July 31, 1992). This rule would specify disinfection treatment techniques, reporting requirements, and performance-monitoring protocols for subject PWSs, and would identify specific microbial organisms of concern. Though a proposed rule was originally expected in August of 1995, it had been postponed indefinitely as this text went to press.

Finally, EPA has proposed new maximum residual disinfection levels (MRDL) for disinfection by-products (Federal Register, July 29, 1994). The NPDWRs for these MRDLs arise out of EPA's 1991 DWPL, which listed disinfection by-products and inorganic chemicals as contaminants of concern. Once finalized, PWS operators will need to comply with these MRDLs in the same manner as any other MCL.

Secondary Standards

In contrast to NPDWRs, National Secondary Drinking Water Regulations (NSDWRs) are not federally enforceable, but serve instead as guidelines for the states. NSDWRs are intended to address the aesthetic qualities of drinking water. Accordingly, EPA has established secondary maximum contaminant levels (SMCLs) for contaminants that may adversely affect the odor or appearance of drinking water, specifically: aluminum, chloride, color, copper, corrosivity, fluoride, foaming agents, iron, manganese, odor, pH, silver, sulfate, total dissolved solids (TDS), and zinc (EPA, § 143.3). States are permitted to establish levels higher or lower than those established by EPA provided that public health and welfare are not adversely affected.

Monitoring and Notification Requirements

NPDWRs require PWSs to regularly monitor the composition of their drinking water for the presence of a number of designated chemicals and to report these results to their state water-quality agency or EPA (EPA, §§ 141.21-141.30). Monitoring frequency and detail depend on the size of the PWS and the level of contamination. Affected health and education facilities should contact their state water agency to determine their monitoring schedule or to request a waiver, which is generally available on a contaminant-specific basis.

NPDWRs contain various reporting and notification requirements (EPA, § 141.31 et seq.). For example, water suppliers must report to the state within 48 hours of the failure to comply with any NPDWR. If an MCL or treatment standard has been exceeded or violated or if the PWS has failed to comply with required monitoring or testing procedures, a PWS must publicly notify both its customers and the supervising state regulatory agency (EPA, § 141.32). Specifically, a PWS must:

1. notify its local newspaper within 14 days of the occurrence, and
2. notify each customer in writing within 45 days of the occurrence (unless the violation is corrected prior to that time)

The notice must clearly explain the violation, the measures required to correct it, and any potential adverse health effects (USC, § 300g-3(c)). For exceedances or violations that pose an acute risk to public health, the PWS must report the occurrence to local radio and television stations within 72 hours.

Exemptions

Under the SDWA, a state may exempt a PWS from any MCL or treatment technique requirement if it finds that:

1. due to compelling factors (which may include economic considerations), the system is unable to comply
2. the system was already in operation on the day that the MCL or treatment technique in question went into effect or, for new systems, that no reasonable alternative source of drinking water is available to the system; and
3. the exemption will not result in an unreasonable risk to health (USC, § 300g-5)

At the same time it grants an exemption, the state must prescribe a compliance schedule and any required interim-control measures.

For exemptions from an NPDWR promulgated after June 19, 1986, the system's final compliance date with the MCL or technique generally must be within one year of when the exemption is issued. The state may extend this date for up to three years. For systems with 500 or fewer service connections that need financial assistance to comply with their MCLs, the state may renew the exemption for additional two-year periods if the system is taking all practicable steps to comply with the provision.

Variances

Variances from NPDWRs may be granted to a PWS if it:

1. is already applying BAT to its water supply *and* is unable to meet an MCL because of the characteristics of its raw water source; or
2. can demonstrate that the nature of the raw water source makes a specified contaminant-treatment technique unnecessary to protect human health (USC, §300(g-4))

Variances will not be granted if they will create an unreasonable risk to public health. In addition, variances cannot take effect until there has been an opportunity for public notice and a hearing. As with exemptions, when a state grants a variance, it must prescribe a schedule for compliance with the NPDWR and the implementation of any necessary control measures.

Variances and exemptions from the MCL for total coliforms, or variances from any treatment requirement specified in 40 C.F.R. §§ 141.70-141.75 (part of the SWTR) will not be granted (EPA, § 141.4(a)).

14.4   LEAD CONTAMINATION IN SCHOOL DRINKING WATER

Lead occurs in drinking water from two sources: (1) lead in raw water supplies and (2) corrosion from plumbing materials in the water-distribution system. According to EPA, most lead contamination results from corrosion by-products (Federal Register, June 7, 1991). A 1991 EPA rule (the "Lead and Copper" Rule) established an NPDWR setting forth required treatment techniques for controlling lead and copper levels in public water systems (EPA, § 141.80 et seq.). These requirements apply if these contaminants exceed levels of 0.015 milligrams per liter (mg/L) for lead and 1.3 mg/L for copper in more than 10% of targeted tap samples. The Lead and Copper Rule also requires monitoring, public education, and service-line replacement. Most of these requirements apply both to CWSs and NTNCWSs.

In addition to the Lead and Copper Rule, the SDWA contains other provisions addressing lead contamination in drinking water. Section 1417 of the SDWA prohibits the use of lead solder or flux (solder or flux containing more than 0.2% lead) and lead-bearing pipes and fittings (those containing more than 8% lead) (U.S.C. § 300g-6). This section also imposes special public-notification requirements. Specifically, PWSs must identify and notify those persons who may be affected by lead contamination in their drinking water, if such contamination is either (1) the result of lead in the PWS's construction materials or (2) caused by corrosivity of the water supply sufficient to cause leaching of lead. PWSs were required to provide these customers with a one-time notice by June 19, 1988.

Under section 1464 of the SDWA, EPA is required to distribute to the states a list of each brand and model of drinking-water cooler identified by EPA as not "lead-free" (U.S.C. § 300(j)). In a "lead-free" water cooler, no part or component that may come in contact with the drinking water may contain more than 8% lead. In addition, if a cooler contains solder, flux, or a storage tank interior surface containing more than 0.2% lead which may contact drinking water, it is not considered "lead-free."

The SDWA requires EPA to publish a guidance document and a testing protocol designed to assist schools in determining the source and degree of lead contamination in their drinking-water supplies and how to remedy such contamination. Each state must disseminate these documents—as well as a list of non-"lead-free" drinking-water coolers—to local educational agencies, private nonprofit elementary and secondary schools, and day-care centers. States may also establish a program to assist local educational agencies in testing for and remedying lead contamination at their schools. EPA is authorized to provide states with grants for these purposes. Schools should contact EPA or the appropriate state agency for additional information about lead in school drinking water.

14.5   GROUNDWATER PROTECTION

In recent years, Congress has empowered EPA to protect groundwater more vigorously. Leaking underground storage tanks (LUSTs) that contain solvents and gasoline products, ground injection of wastes, and runoff from agricultural, forestry, municipal, and residential operations have all contributed to the contamination of groundwater.

Wellhead Protection Program

A wellhead protection area is the area, both above and below ground, surrounding a well or wells used to supply a public drinking-water system (USC, § 300h-7(e)). Each state is primarily responsible for developing and implementing its own wellhead protection plan (WPP). No state, however, is under any obligation to develop a WPP, and if a state does not develop a program, there is no federal structure waiting to fill in the gaps. Thus, the protections and burdens of each WPP are left almost entirely to each state's discretion.

For those states that do decide to develop a program, each WPP must map and define the surface and subsurface areas of those PWS wells that are thought to be threatened by contamination. The program must also identify artificial sources of contamination that could adulterate PWS drinking-water supplies (USC, § 300h-7(a)). Once these surveys are completed, the WPP must set forth recommended educational, protective, and remedial measures, including cleanup regimes and identification of alternative sources of drinking water.

States can apply for federal funding for their WPP development and implementation activities. Those states that institute WPPs must have them reviewed and approved by EPA to ensure that they are adequate to protect PWSs. A WPP is deemed "approved" if the EPA administrator does not take action on it within nine months of receipt (USC, § 300h-7(d)). WPPs have no stated effect on either property owners' water rights or a state's right to protect its water supplies. Citizens cannot use SDWA's citizen-suit provisions, discussed below, to compel their state to implement a WPP.

Critical Aquifer Protection

State and local governments may be eligible for federal funds if one of their aquifers qualifies as a critical aquifer protection area (CAPA) within a Sole Source Aquifer (SSA)[1] (USC, § 300h-6). While EPA has the authority to initiate SSA designations, it has made a policy of acting only in response to outside petitions.[2] Federal funding may not be used to fund projects that may endanger these aquifers, and SSAs are favored for inclusion in CAPA demonstration programs. CAPA funding must be used to implement an aquifer-protection demonstration program known as a comprehensive management plan (CMP) (USC, § 300h-

---

[1] SSAs are aquifers that supply drinking water for at least 50% of the population in cities with an alternative supply.

[2] Petitions may be submitted to EPA by any individual or organization and must address procedures and criteria outlined in EPA's "Sole Source Aquifer Designation Petitioner Guidance" (EPA 440/6-87-003).

6(f)). The CMP must identify existing and potential contamination sources, and recommend specific management practices designed to prevent and/or remediate contamination of the CAPA.

Both CMP and section 208 Sole Source Aquifer (SSA) protection plans can effectively serve as wellhead protection plans (WPP), discussed in the preceding section. These may be of interest to health and education facilities that either have themselves threatened or contaminated local aquifers by their past practices or, alternatively, wish to ensure that their current groundwater sources are protected from outside contamination.

Underground Injection Control (UIC)

EPA retains authority under the SDWA to enforce the provisions of a state's WPP and to bring enforcement actions, typically referred to as "UIC administrative orders," for violations of UIC underground-disposal requirements. Most of the regulations that detail the specifics of deep-well-injection programs are found in the Resource Conservation and Recovery Act (RCRA) (EPA, § 265).

14.6   ENFORCEMENT

Section 1414 of the SDWA provides EPA with a variety of traditional enforcement mechanisms (USC, § 300g-3). For example, EPA may bring a civil action in federal district court to require compliance with a NPDWR. In addition, if EPA determines that violations of the SDWA's monitoring or public notification of violation provisions have occurred, the Agency will notify the PWS and the state, allowing the appropriate agency 30 days to take action. If the state has not begun appropriate enforcement action, EPA must take action, and may seek civil or administrative fines from the PWS of $25,000 and $5,000 per violation, respectively.

EPA may also issue compliance orders if the PWS and/or the state regulatory agency fail to act within 30 days of an MCL, variance, or exemption violation. EPA may issue such compliance orders after an opportunity for public notice and a hearing (USC, § 300g-3(a)). The Agency may assess a $5,000 administrative fine per violation for failure to comply with such orders. EPA may also choose to proceed judicially, pursuing injunctive relief and fines of up to $25,000 per day of violation. Finally, if the PWS's water presents an "imminent and substantial" danger to public health, EPA may take immediate enforcement action (USC, § 300(i)).

While the SDWA does authorize citizens' suits for violations of the Act, in contrast to the Clean Water Act's (CWA) citizen-suit provisions, it does not provide for a plaintiff's recovery of monetary penalties (USC, § 300j-8). Thus, the frequency of SDWA suits against PWSs and state-enforcement agencies is far lower than for other environmental statutes with citizen-suit provisions. Attorneys fees, however, are available for SDWA citizen suits.

REFERENCES

40 Code of Federal Regulations § 141.2.
40 Code of Federal Regulations § 141.3.
40 Code of Federal Regulations § 141.4(a).
40 Code of Federal Regulations § 141.4(b).
40 Code of Federal Regulations §§ 141.21–141.30.
40 Code of Federal Regulations § 141.31.
40 Code of Federal Regulations § 141.32.
40 Code of Federal Regulations § 141.70.
40 Code of Federal Regulations § 141.80.
40 Code of Federal Regulations § 142.
40 Code of Federal Regulations Part 142 Subpart B.
40 Code of Federal Regulations § 143.3.
40 Code of Federal Regulations § 265.
FDA.1995. Code of Federal Regulations § 103.35.
56 Federal Register 1470, January 14, 1991.
56 Federal Register 26460, June 7, 1991.
57 Federal Register 33960, July 31, 1992.
59 Federal Register 38668, July 29, 1994.
21 United States Code § 349.
42 United States Code § 300f.
42 United States Code § 300g.
42 United States Code § 300g-1.
42 United States Code § 300g-1(b).
42 United States Code § 300g-1(b)(4).
42 United States Code § 300g-3.
42 United States Code § 300g-3(a).
42 United States Code § 300g-3(c).
42 United States Code § 300g-4.
42 United States Code § 300g-5.
42 United States Code § 300h-6.
42 United States Code § 300h-6(f).
42 United States Code § 300h-7.

42 United States Code § 300h-7(a).
42 United States Code § 300h-7(e).
42 United States Code § 300i.
42 United States Code § 300j-8.
42 United States Code § 300j-24.

Chapter 15

What Happens in Siting/Zoning Environmental Conflicts?

15.1   SAFETY ISSUES IN ZONING AND LOCAL PERMITS

Institutions must change and grow, and when this growth takes the form of new construction, the safety and environmental manager will be a key contributor to the clearances that are required. Zoning is the process by which uses of land are regulated by public bodies in a planned, rational manner (McQuillin 1987). Zoning regulates both public and private uses of property for the greater good of the community. Beyond that textbook definition, zoning is a topic that may draw "intense, heated debate" (*Laurel Heights* v *University of California* 1987, 453) whenever specific projects are proposed for specific sites. Universities, schools, and hospitals are among the institutions affected by local zoning controls (Olivieri 1975).

When an institution is surrounded by residential neighborhoods, loud opposition to changes in zoning, or variances from an existing zoning plan, will be a frequent occurrence. Georgetown University's lengthy dispute to site its power plant (*Citizens Assn. v DC Board of Zoning Adjustment* 1976) and Boston University's unsuccessful rezoning effort illustrate the barriers that organized opposition from neighbors can impose (*Bedford* v *Trustees of Boston University* 1988). The University of California at San Francisco's unsuccessful, much-litigated struggle, which consumed a decade after the initial decision to use residential area buildings for biomedical research was made, seems like every campus planner's worst nightmare. The proposal was overwhelmed by more than 5,000 pages of adverse comments on the environmental effects and after several trips to the state's Supreme Court (*Laurel Heights* v *University of California* 1993).

Why should neighbors enter into conflicts with institutions? Individuals who own homes or smaller apartment houses perceive themselves as investors whose capital is tied up in the resale value of the residence. They perceive some expansion or construction actions as lessening resale value. Even the lights used in a building can be criticized as an environmentally adverse effect, though courts have held the effects of normal lighting to be insignificant. (*Laurel Heights* v *University of California* 1993). These are perceived to be a threat not just to lifestyle but to financial well-being. Much has to do with perceptions and relations among neighbors. Their perception of your institution may be of a big, secretive bully whose burdens are borne by nearby residents. If it is, you personally may become a target of anger in the passionate airing of objections at zoning hearings.

Zoning projects require preparations greater than just architectural plans. Successful relations with the institution's residential neighbors usually take the form of advance discussions with elected officials and local "thought leaders," visits with community groups to discuss the site plan, door-to-door visits to answer questions and leave literature, and in severe cases, future purchase-value guarantees or other financial arrangements to protect the homeowner's property value, in return for homeowner support of the zoning action. This latter option gives a promise to purchase the home at existing comparable prices, prior to the zone change, if the owner decides in the next X years to sell and then finds that the value is lessened by the construction or expansion. This form of contingency insurance for home values is an expensive offer, so it is not used in any but the most sensitive situations.

When opponents speak against expansion or new construction at a university, hospital, or school, safety threats to neighbors may be an issue raised in the petitions or complaints. Biomedical research carries the stigma of potential "escape" of some harmful research material, made worse if opponents discern concealment of the research purpose for the new institutional addition (*Laurel Heights* v *University of California* 1988). Typically, the institution's safety manager will be in the middle of a very public debate, so this section explores some of the likely controversies and responses.

Traffic is both a noise and a safety issue in zoning disputes (*Laurel Heights* v *University of California* 1993). Many of the university zoning disputes have concerned excessive car and truck traffic on residential streets (*Marjorie Webster Junior College* v *DC Board of Zoning* 1973; Appeal of Community College 1976). Opponents may come armed with trip-reduction demands, aided by the recent enactments of Clean Air Act amendments (Pub. L. 101-549, 1990) and the Intermodal Surface Transportation Act. The response to challenges concerning traffic flow will vary: some institutions may adopt ride-share or other traffic-avoidance schemes; some will study lane closures, redirections of flow, and other physical adaptations, and show that traffic capacity of local streets may be sufficient. In some cases, off-street parking shortages could be improved by the institution's garage that will be part of the new site.

For universities, creation of additional parking facilities and limits on resident student use of cars are typical responses to the traffic complaints of area residents. The university could do all these measures and might also cooperate with residents to petition for more parking enforcement on the affected residential streets. For hospitals, construction of parking decks and sound barriers for areas around emergency entrances could be offered.

Fire-code aspects of the new building are also part of the zoning opponents' arguments. Past records of fire-inspection violations are a basis for another possible objection. This becomes an issue if sophisticated neighbors hire experts to challenge the adequacy of fire lanes, evacuation capabilities, and/or the proximity of the institution's storage tanks to the nearest residences. For example, a hospital's aboveground storage tank for a flammable oxygen gas might trigger objections because of fears regarding ruptures if sparks are generated when the tank is struck by a delivery truck. Safety managers working with local fire-prevention teams can show on paper that all applicable safety standards have been met, and that hazardous-materials truck-loading safeguards have been established to prevent leaks or spills. Lawyers for the institution will be sure to avoid a blanket promise that the spill could *never* happen, but reasonable precautions explained in a rational fashion should be as persuasive as one can get.

The officially permitted "use" of a building like a school or hospital is governed by the local zoning code. A university that declares a use, and later is shown to intend a use that involves greater risk, is vulnerable to attack for both risk and the concealment of relevant information from the interested public (*Laurel Heights* v *University of California* 1987, 455). Some zoning "uses" will be changed by conversion of buildings, and reasonable restrictions can be imposed by zoning officials (*Trustees of Tufts* v *City of Medford* 1993). There may be disputes concerning the safety for instructional use of a building that was designed for light industrial use, for example. If a hospital converts an adjacent storage building into outpatient clinics, disputes about that use in a certain zone may be more difficult, if the new use of a site carries with it complexities of parking, fire exits, and so on. The zoning authority may use its power to impose extra requirements to minimize adverse impacts of the proposed use (*Cornell University* v *Bagnardi* 1986). The exception will be a state-owned institution that has statutory exemption from local zoning laws (*Inspector of Buildings* v *Salem State College* 1989; *City of Newark* v *University of Delaware* 1973). Not all state institutions are so exempt, and those that avoid a zoning requirement may face a comparable scrutiny in the environmental-impact process (*Laurel Heights* v *University of California* 1987, 455).

15.2 ENVIRONMENTAL ISSUES IN ZONING AND LOCAL PERMITS

The conventional issues of increased traffic and noise are issues that zoning cases routinely examine. There are likely to be few occasions when a county or town zoning board hears the more exotic controversies about radiation, hazardous-waste on-site storage, and laboratory chemicals. Novelty of an issue engenders caution among the unprepared listeners. The university or hospital that presents these more complex issues to zoning officials should not assume that the municipal officials will be sympathetic. A smaller community's elected planning board might be unable or unwilling to manage the controversies that are raised by the opposition. A prudent institution will form a proactive team, composed of the site safety manager for the institution, the zoning lawyer or other advisor, and the public-relations manager.

Three typical issues needing advance explanation, education, and defensive work are radiation, additional pollution loading of public facilities, and chemical-waste storage.

Radiation standards are set by the federal Nuclear Regulatory Commission (NRC) for storage and disposal of the low-level radioactive waste that a hospital or university routinely produces (NRC 1995). These federal regulations are very specific and conservative. The standards apply uniformly across the country. The NRC rules have "preemptive" effect, so that they override local and state rules that are different or that add extra layers of controls. The city zoning officials should not be allowed to upset the federal norms that apply consistently across all private and public entities. A local sewer agency, for example, cannot set a standard for the water effluent from radiation-treatment rinse water that is more stringent than the NRC levels of permissible radioactivity in the wastewater stream from a hospital's radiology department. The local standards for suspended solids, biological oxygen demand, and other parameters for wastewater content are not affected by the federal law's preemption of any local control of the radioactive effluent. The zoning board cannot impose a sewer standard as a condition on use of the site that differs from the federal standard.

Concerns about additional burdens of the new facility upon local environmental-control plants or facilities can be expected as well. If the opposition is sophisticated enough to advocate for the appealing benefit of "source reductions" when the new "source" is proposed, the claim might be made that pollution-prevention goals require the local zoning board to limit amounts of pollutants by means of numerical levels, written into stipulations entered in the zoning agreements. The prudent institution will be wary of such controls; the risk of being artificially "capped" against growth, at the very outset of the construction and operation of a new facility, should be recognized by the hospital or university.

The institution should expect that its safety manager and the design engineers will be aware of any local pretreatment thresholds for the publicly owned treatment works (POTW). If the effluent projected to flow from the new site is within the marginal additional capacity of the local POTW, but challenges to the zoning based on "sewer overload" are heard, then a letter of nonobjection should be obtained from the sewer officials. Air-pollution controls are outside zoning control and are discussed in Chapter 5. For both types of environmental-control programs, the opposition's demand for a pollution-prevention agreement should raise red flags for the safety manager.

The level of management that gets involved in zoning discussions should be carefully considered. Assume that a managing official wishes to win over a hostile group of neighbors; they ask her to sign a promise of a pollution-prevention program for the new facility and in a burst of good feelings she agrees. The fine print must be read by engineering managers before a "deal" is announced. Public-relations managers and general managers should be prepared to deal with such pollution-prevention claims broadly and with an eye to feasible implementation of reasonable promises. A fair dose of skepticism is advised in order to avoid the error of the institution's managers signing on to an artificially low ceiling of emissions or effluents. It is possible that a zoning agreement can be a win/win environmental outcome; but signing up too quickly to a bargain that cramps the institution can spoil the intended use of the new facility before it opens.

Chemical-waste hazards are relevant to a zoning decision, when a well-informed opponent seeks to attack on-site storage plans. It is virtually inevitable that a large institution will be a "generator" of hazardous waste (EPA 1990). Chapter 8 of this text explains the issues in great detail. Medical gases and propane may be required by NFPA fire-code standards to be isolated in certain areas on the grounds, and the local building code may incorporate by reference the NFPA standard. Radiation equipment for medical research excites fear among neighbors who do not understand its safeguards and controls (*Laurel Heights* v *University of California* 1987). The safety manager will be a supporting player in the institution's zoning team, and should reexamine the fire-code standards that will apply as the architect is developing the plans that will be submitted to zoning officials.

Not all challengers are residential homeowners. Sometimes, objections to a new site may come from interested entities that use safety as a basis of objections. Competing hospitals argued in a Maryland case that a new hospital should not be located around the runway of an airport, but their objection was unavailing (*Clinton Community Hospital* v *So. Maryland Medical Center* 1975).

15.3   OPPOSITION TO ZONING CHANGES

Fear of the unknown, such as the uninformed person's skeptical view of the safety of labs doing biotechnology and genetic research, can play a role in neighbors' opposition to zoning permission for a research or engineering facility.

The case of university medical-research laboratories at the University of California at San Francisco illustrates a very harsh lesson about zoning (*Laurel Heights* v *University of California* 1987, 1988, 1993). Buildings that were vacated by a relocating insurance company were made available to the university, which intended to conduct biomedical research work in the buildings. Opposition from residents was strong because the benign office workers who were the former tenants would be replaced by high-tech medical researchers working with possibly harmful substances in their neighborhood (*Laurel Heights* v *University of California* 1987). San Francisco's sensitivity to environmental issues and the news-media attention to the controversy made for a turbulent set of court conflicts. Ten years after the dispute began, no medical research has been done and the site is largely vacant.

Methods for anticipating and avoiding trouble are focused on early and consistent communications. Do not risk the backlash that allegations of concealed purposes would engender (*Laurel Heights* v *University of California* 1987). The incoming institution can invite community leaders to visit a comparable facility, at the expense of the institution. The institution can describe safeguards and limits that it voluntarily is using. A publicized community "open house" weekend with a mailing of invitations can be used as a means of showing the beneficial purpose of the new facility and the reasons for the zoning change. Credible third-party safety experts such as the fire chief and health-department managers can be given a tour of the facility or can examine the plans, with hopes for an endorsement of the reasonableness of the entity's claims. The home-address lists of staff and employees of the institution should be screened so that those who live in the area are given a set of brochures about the project with a request that they share the information with neighbors. In this way, multiple sources of both "peer review" and "expert evaluation" can support the entity's claim that its zoning change will be safe for neighbors.

The entity's track record of environmental compliance in other sites can be expected to receive scrutiny from opponents, so the safety manager will need to be ready to explain why the new or expanded site will not have the same concerns. The public record of complaints and how they were handled by the institution is a part of the "record" on which opposition will be based. The safety manager's position makes this manager very well positioned to turn the negative aspect of past conduct into a beneficial reason as to why the newer facility will be a justifiable improvement.

Cancer is a much-feared word. If the new facility is a cancer research or treatment center, heavier opposition is anticipatable (*Laurel Heights* v *University of California* 1987). Some portion of the institution's public-communications materials should emphasize that the work benefits public health; the materials to be used are well controlled; and the facility is investing in safety devices to reassure neighbors that the laboratory will not spread cancer-causing materials into the area surrounding the site. Concepts like containment and high-efficiency exhaust filtration are well known to the technical community; the smart safety manager makes certain that public-relations and government-relations staff are fully briefed about the probable safety issues. A straightforward explanation about medical research in virology, and of the regulatory constraints on handling pathogens, can dispel ignorant claims that strange viruses will endanger children and pets, for example. Low-level radioactive waste bears a stigma that alarms neighbors (*Laurel Heights* v *University of California* 1987), yet lifesaving medical therapeutics often depend on diagnostic radiolabeled pharmaceuticals.

15.4   ENVIRONMENTAL-IMPACT STATEMENTS AND NEGOTIATIONS

Opponents who can delay a project can sometimes kill it by frustrating its intent or its funding sources. The San Francisco biomedical research labs discussed above were not installed in the residential neighborhood, even after ten years of conflict, because the litigation over environmental-impact assessment documents stretched out longer than anyone could have anticipated (*Laurel Heights* v *University of California* 1987, 1988, 1993). Demands for a federally funded or licensed project to produce an "environmental impact statement" (EIS), or its state equivalents for nonfederal projects (*Laurel Heights* v *University of California* 1988, Sax 1972), serve as a means of delay and litigation.

The process of preparing an assessment and impact document that is more than merely a "cursory" one (*Laurel Heights* v *University of California* 1987) will consume several years for any major opposed facility. This delay covers both the impact statement's research and preparation time, and the court delays as opponents challenge the adequacy of the agency decisions whether to perform an EIS and then what elements to include in the assessment (*Laurel Heights* v *University of California* 1993). Because many construction projects receive federal funding or require federal licensing (*Calvert Cliffs* v *U.S. Energy Comm.* 1979), the EIS process often applies when an education or health-care facility expands. The aspect of the EIS content that will cause the most controversy is the required search for, and discussion of, alternatives to the proposed action. Opponents can always think of a better alternative outside of their neighborhood and the EIS drafting team will spend a lot of what will be perceived as wasted effort in pursuit of a suggested alternative before disqualifying that less-desirable option.

In the federal National Environmental Policy Act (NEPA) of 1970, Congress acted with a rhetorical flourish to help the environment, an unopposable objective, but it did so without really considering the degree to which it was creating a future impediment to all federal actions. NEPA requires federal agency "actions" (unless exempted) to have an environmental assessment, and then either a Finding of No Significant Impact or a full Environmental Impact Statement (CEQ

Part 1500). Impact includes the psyschological fear of accidents at the new facility (*Metropolitan Edison* v *People Against Nuclear Energy* 1983).

Sophisticated opponents will routinely argue that the full EIS is needed and once it is done, will further argue that the EIS is inadequate (*Laurel Heights* v *University of California* 1988). NEPA itself does not forbid construction of federal buildings or projects (*Stryker's Bay* v *Karlen* 1980), but its expense and the utility to opponents is a significant factor in slowing down the steps that an agency considers to be "progress" (Menell and Stewart 1994). In 1991, for example, there were 456 environmental impact statements and 94 new NEPA lawsuits. (Menell and Stewart 1994, 1023)

State facilities are in some cases also covered by state environmental quality laws (*Laurel Heights* v *University of California* 1988, Sax 1972). State-funded programs will be covered by these laws, and in the extreme situations, facilities built with the use of publicly financed bonds or grants will also require a state EIS.

When a university acts as a federal contractor, it is deemed to be part of an "agency" for purposes of the NEPA requirement to prepare an environmental impact statement. This means the university can be enjoined from operating a federally funded program without an impact statement (*Foundation* v *Heckler* 1986).

REFERENCES

Appeal of Community College of Delaware County, 254 A.2d 372 (D.C. App. 1976).

Bedford v *Trustees of Boston University*, 518 N.E.2d 874 (Mass. App. 1988).

Calvert Cliffs Coordinating Committee Inc v *U.S. Atomic Energy Commission*, 449 F2d 1109 (DC Cir., 1971).

CEQ, Implementing Regulations of NEPA by the Council on Environmental Quality, 40 CFR Parts 1500 et seq.

Citizens Assn. v *District of Columbia Board of Zoning Adjustment*, 365 A.2d 372 (D.C. App. 1976).

City of Newark v *University of Delaware*, 304 A.2d 347 (Del. Ct. Ch. 1973).

Clinton Community Hospital Corp. v *Southern Maryland Medical Center*, 510 F.2d 1037 (4th Cir., 1975), cert. den. 422 U.S. 1048 (1975).

Cornell University v *Bagnardi* 510 N.Y.S.2d 861 (1986).

Foundation on Economic Trends v *Heckler*, 756 F.2d 143 (D.C. Cir., 1986).

Inspector of Buildings v *Salem State College*, 546 N.E.2d 388 (Mass. App. 1989), rev. den. 548 N.E.2d 887 (Mass. 1990).

Laurel Heights Improvement Assn. of San Francisco Inc. v *Regents of the*

University of California, 193 Cal. App.3d 467, 238 Cal. Rptr. 451 (1st Dist. 1987).

Laurel Heights Improvement Assn. of San Francisco Inc. v *Regents of the University of California,* 47 Cal. 3d 376, 764 P.2d 278, 253 Cal. Rptr. 426 (1988).

Laurel Heights Improvement Assn. of San Francisco Inc. v *Regents of the University of California,* 6 Cal. 4th 1112, 864 P2d 502, 26 Cal. Rptr. 2d 231 (1993).

Marjorie Webster Junior College v *District of Columbia Board of Zoning Adjustment,* 309 A.2d 314 (D.C. App. 1973).

McQuillin, Eugene. 1987. Law of Municipal Corporations 3d Ed. Ch. 25.

Menell, Peter, and Richard Stewart. 1994. *Environmental Law and Policy.* Boston: Little Brown.

Metropolitan Edison v *People Against Nuclear Energy,* 460 U.S. 766 (1983).

National Environmental Policy Act, 42 USC §§ 4321-4347, Pub. L. 91–190, 83 Stat. 852 (1970), implemented by 40 CFR Parts 1500 et seq.

Note. 1983. Psychological Effects at NEPA's Threshold. Columbia L. Rev. 83: 336.

NRC-Nuclear Regulatory Commission. 1995. 10 Code of Federal Regulations parts 20 et seq.

Olivieri, Daniel. 1975. Zoning Regulations as Applied to Colleges, Universities, or Similar Institutions for Higher Education. *American Law Reports* 3d 64: 1138.

Sax, Joseph. 1972. Michigan's Environmental Protection Act of 1970: A Progress Report. Michigan L. Rev. 70: 1003.

Stryker's Bay Neigborhood Council Inc v *Karlen,* 444 U.S. 223 (1980).

Trustees of Tufts College v *City of Medford,* 616 N.E.2d 433 (Mass. 1993).

Chapter 16

Managing Compliance for an Educational or Health-Care Institution

16.1 INTRODUCTION

The field of safety and environmental protection has undergone a vast transformation in the past 25 years. The early 1970s saw both federal and state environmental and safety legislation enacted in attempt to control pollution that not only affected the environment but public health as well. This legislation described specific regulatory standards and enforcement actions to be used to police the environment. Its funding and its enforcement goals were industrial in orientation and did not address the nonprofit health-care or educational sectors of the economy.

Although much of this "command and control" environmental legislation was very specific in nature, many of the implementing regulations were continually challenged in the courts by both industry and public-interest groups. Usually these EPA or OSHA rules, guidance documents, and interpretations of statutes were industry-focused and were either nonapplicable or very difficult to implement successfully in an institutional setting. The evolution of these regulations has resulted in increased demands on small staffs and limited budgets of institutional settings.

As a result, prudent management of the institutional environmental function has evolved into a very complex task.

16.2 KEY ELEMENTS

Since every institution is different, every occupational-health and environmental-safety program will be different also. However, there are key elements of a successful program that are the same or very similar in all programs:

- top management support
- good management of the function
- effective staffing
- effective hazard-surveillance reporting
- effective preventive programs and procedures
- ongoing communication

This chapter will address the elements of success and how they can be achieved.

16.3   WINNING MANAGEMENT SUPPORT

Many times, increased administrative support from the top of an institution in support of the safety function comes about as a result of some type of catastrophe related to occupational health and environmental safety that has befallen the institution. The money becomes available after the need has been tragically demonstrated, after a period of short budgets and long risks.

Institutions vary but the belated support for safety is a chronic problem facing managers. The analogy to the world chemical industry serves as a prime example of this scenario. Union Carbide Corporation is a leader in its support and funding for occupational-health and environmental-safety programs in large part due to adverse publicity from past traumatic environmental experiences. Physical-safety lessons also seem to be taught by publicized failures. The 1958 fire at the Our Lady of the Angels School in Chicago with its 93 fatalities resulted in a nationwide increased awareness and response to improving life safety in institutional settings.

Recently, several universities have been dealing with EPA enforcement actions that have resulted in hundreds of thousands of dollars in monetary fines. The institutions' increased awareness in dealing with these problems have resulted in upgrading their occupational-health and environmental-safety programs. It takes prudent managers and courage to make positive change *before* the disaster proves the need for change.

16.4   AVOIDANCE OF EXPENSIVE PENALTIES

Regulatory agency monetary penalties can be quite costly. OSHA is one example. During the 1990 budget negotiations, Congress, in its search for new sources of money, changed the OSHA penalty structure, partly in response to a small but vocal minority that wanted to increase OSHA's enforcement penalties to create more pressure on business. Congress included an amendment in the budget package which increased OSHA's penalty structure by a factor of seven. However, Congress also turned the Agency into a revenue source, charging it with increasing its revenues from penalties from the then-current level of $40 million per year to a total of $900 million over five years. In the 20 years of OSHA's existence, total fines have been less that $4 billion, with approximately $90 million in penalties assessed in 1991. Collected penalties have always been less than those proposed, often by half. For example, OSHA penalties include:

Willful violations
Maximum $70,000   Minimum $5,000
Repeated violation
Maximum $70,000
Serious and other than serious
Maximum $7,000
Failure to abate for each calendar day beyond abatement date
Maximum $7,000
Failure to post OSHA notice $1,000
Posting of OSHA 200 summary $1,000
Failure to post OSHA citation $3,000
Failure to maintain OSHA 200 injury log $1,000
Failure to report fatality/catastrophe $5,000
Failure of access to records $1,000
Assaulting compliance officer $2,000

The EPA will likewise impose penalties for the following violations:

Resource Conservation and Recovery Act
Civil $25,000/day/violation
Criminal $50,000/day/violation

Clean Air Act
Civil Penalties
     Minor volation/field citation $5,000/day/violation
     Administrative assessment $25,000/day/violation
     Criminal penalties
     Knowingly violate (person) $250,000/day/violation/
                                       and up to 5 years prison
     Knowingly violate (person) $500,000/day/violation

CERCLA
     Civil Class I $25,000/day/violation
     Civil Class II $25,000/day/violation
     Subsequent Civil Class II $75,000/day/violation

The Department of Transportion will enforce penalties under the provisions dealing with hazardous-materials transport:

HMTA (Criminal) $25,000/viol./5 yrs.
HMTA (Civil) $10,000/violation

16.5   WINNING POLICY SUPPORT AND COMMUNICATION

Senior management of the institution should communicate and define expectations for the institution's occupational-health and environmental-safety organization. A basic environmental, health, and safety policy should be issued from the chief executive officer of the institution and include a commitment for full implementation. This policy should ensure compliance with all relevant environmental, health, and safety laws and regulations.

The policy should be proactive and call for appropriate measures to protect the health and safety of students, patients, faculty, staff, and visitors by either eliminating where possible or limiting to the lowest practicable levels, any adverse effects to human health and the environment of exposures at the institution.

In such a policy statement, all supervisory personnel should be made responsible for protecting the health and safety of students, employees, and patients under their supervision and care. All levels of management, and particularly first-line supervisors, have a special responsibility for the safety of their employees and own work area. They are familiar with their own areas, know their people, their operations, and have operational authority and control. The policy should instill a positive attitude toward accident prevention in the institutional members and encourage the use of sound safety practices in their professional and community activities.

Mangement commitment should be made visible through participation in committees, reporting lines, planning and implementation of long-range institutional goals, and ensuring that the occupational-health and environmental-safety group has the necessary resources to perform their jobs in an proactive, effective manner.

16.6   OVERCOMING INTRA-INSTITUTION RIVALRIES

A typical institution is made of many individuals with various job functions. Administrators, doctors, researchers, housekeepers, maintenance workers, office staff, and so on each rule over their own little (or big) kingdom. It is human nature for people to be protective of their turf. The institutional setting is no exception, and indeed the presence of tenure and of limitations on internal discipline of some groups makes for a more difficult control situation.

This intra-institutional arrangement can be an insurmountable barrier to implementing new safety and health policies for compliance with the many new laws and regulations that are continually evolving. Responsibilities should be clearly defined by senior managers. The stronger the support is from the top, the easier it is to get past those turf barriers to implement new policies and procedures that will lead to an effective occupational-health and environmental-safety program.

16.7   EFFECTIVE MANAGEMENT

It is important that the individual with direct supervision over the environmental, health, and safety program have a wide range of talents and abilities. Whether it is a small institution where the management of the environmental, health, and safety program is someone's collateral duty, or a large institution where the manager oversees a large staff of professional and support staff, it is important that the manager be able to handle a wide range of issues effectively. Diplomat, mechanic, regulatory expert, and team-leader skills are needed.

Those individuals with oversight over institutional occupational safety and environmental programs are known by many titles and generally provide mangement over the following types of programs: air quality, water quality, indoor environmental quality, radiation (ionizing and nonionizing), biosafety, industrial hygiene, injury and illness prevention, lead poisoning–prevention programs, asbestos programs, respiratory-protection programs, medical monitoring, ergonomics, chemical safety, fire protection, hazardous-waste management, laboratory safety, transportation safety, food sanitation, accident-investigation and insurance-claims management, workers' compensation, EPA and OSHA regulations.

Typically the individual is asked to ensure that the institution maintains compliance with all relevant state, federal, and local regulations. Individuals in this position are usually asked to demonstrate high-level verbal, writing, training, and management skills. There is no single skill set or degree that prepares one for the complete role, and continuous training and improvement of skills should be expected.

Typical educational requirements include study in fields such as environmental health, industrial hygiene, public health, health physics, and/or other similar areas. Typical salaries range from $40,000 for a small institution to over $140,000 for an institution with a large staff and high visibilty.

The reporting structure varies from institution to institution. Generally, the higher the level of manager to whom the safety professional reports, the more emphasis is placed on the overall program by the institution's hierarchy.

An effective environmental health and safety manager should be able to see the big picture and identify where all the pieces of his/her program fit together. An effective manager should have the same skills described in the following section on effective staffing.

16.8   EFFECTIVE STAFFING

Another important element of a successful program has to do with effective staffing. In today's world of lean staffs and business re-engineering, it is important that each individual hired be as competent as possible. Athough technical skills are important, there are other skills that can make an individual very valuable to an organization.

A good staff member should be able to function independently and have strong organizational and interpersonal skills. As a part of functioning independently, people management and communication skills are important. Today's environmental professional spends a great deal of time interfacing with diverse populations. Written and oral communication skills need to be well developed. Training, explaining results of testing, and responding to emergencies are just a few instances where contact with constituents occurs on a regular basis. Staff members must be able to communicate in an effective manner and leave the public with a sense of confidence in their skills and information, or serious problems can arise. Often, lay people do not have the necessary techinical backgrounds to make knowledgeable judgments about safety and environmental matters. As a result, they depend on information and direction from the environmental professional. Misperception due to miscommunication is a problem that need not occur.

Sales is another important skill in an environmental professional. This skill goes hand in hand with the people and communication skills. We must be able to sell our programs to the constituents in our insistutions. Faculty, doctors, patients, students, office staff, facilities and plant staff, vice presidents, visitors, and so on will need to be interested in their own safety and that of their environment. The professional "sells" behavior change and system improvement to a constituent group or an individual; one cannot expect that alignment with safety or environmental protection goals will be automatic with every member of the university or hospital community.

Individuals working in an institutional occupational-health and environmental program should have a good solid technical background. It is difficult to explain the reason behind a desision based on technical information to someone if you do not understand it yourself. Technical skills are probably more easily learned than the people skills described above. In today's environment of information technology, strong computer skills (word processing, spreadsheet, and database) are certainly valuable assets.

Hiring the right person for this complex task requires more than a simple position announcement. The skills listed above should be communicated with the job title when the position is sent to your personnel department. Personnel recruiters tend to do a much better job of screening when they have some guidelines to follow.

Frequently, staffs can be bolstered by the use of interns, co-op students, and students from apprentice programs. All it takes is a few phone calls to area educational institutions to determine if there are students available to help your program.

The following certification programs are frequently mentioned as qualifications in advertisements for institutional health and safety positions:

Certified Industrial Hygienist (CIH)	Education/experience (5 yr)/exam
Certified Safety Professional (CSP)	Education/experience (4 yr)/exam
Certified Health Physicist (HPS)	Education/experience (6 yr)/exam
Certified Hazardous Material Manager (CHMM)	Education/experience/exam
Certified Environmental Professional (CEP)	Education/Supervisory experience/exam
Registered Environmental Manager (REM)	Education/experience/exam
Registered Environmental Professional (REP)	Education/experience/exam/other approved registration
Certified Environmental Trainer (CET)	Education/experience/exam/extensive training experience
Industrial Hygienist in Training (IHIT)	Education/experience (1 yr)/exam
Associate Safety Professional (ASP)	Education/experience (1 yr)/exam
Occupational Health & Safety Tech (OHST)	Education/experience (5 yr)/exam
Registered Sanitarian (RS)	Education/exam
Registered Hazardous Substance Professional (RHSP)	Education/experience/exam
Certified Hazard Control Manager (CHCM)	Education/experience/exam

Athough having a certification does not ensure that an individual is competent in his/her field, it does indicate that that individual has met a certain minimum set of qualifications related to work in the field.

16.9   EFFECTIVE HAZARD-SURVEILLANCE REPORTING

Early identification and resolution of potential occupational-health and environmental-safety problems is vitally important. Since most institutional organizational structures are decentralized, it is important that primary safety and environmental responsibilities should rest with the working managers. Individuals performing the day-to-day work functions are in the best position to identify potential problem areas.

This principle of subsidiarity teaches that problem solving at the level closest to the need is more likely to be effective than would be a uniform dictation of solutions from above. No matter how large and knowledgeable a safety and environmental staff is, they cannot be everywhere all the time and are most effectively used as a support function for compliance.

There should be a formal reporting structure so that information can be funneled to the individuals who can help solve the problem. There should also be a formal method in place for tracking potential problems. This should also include a procedure for follow-up to ensure that implementation of the corrective action was effective.

16.10   EFFECTIVE PREVENTIVE PROGRAMS AND PROCEDURES

Effective internal procedures and policies should be put into place to ensure compliance with applicable laws and regulations. These procedures and policies should not just be paper props but should be working documents. Training on these procedures and policies should be conducted throughout the organization. What happens in training and what happens in real life should be the same. Periodic audits should be performed to ensure that paper programs and reality match. Programs and procedures have been identified throughout this book that can be used as the basis for effective programs and procedures.

16.11   COMMUNICATION

Communication is an area that has been discussed throughout this book. Effective communication can be used to justify more resources, bolster department credibilty, educate others, and ensure that public perception is accurate.

Committees can be an effective means of communicating between diverse populations. For committees to be effective, the membership should be made up of a diverse group that is really interested in producing results. By the same token, the recommendations should be taken seriously by the top administration officials who should be supporting the group and their mission. Committees should consist of members from different strata in an organization.

16.12   USE OF CONSULTANTS

Consultants should be used when in-house expertise is not available either because of time constraints or budget. Usually, consultants provide specific information for a particular purpose. It is important to determine if the consultant has the necessary experience to perform your job without a long learning curve. Many times when a new regulation needs to be implemented, a consultant is in the same state of confusion as the clients, and does not yet have the necessary expertise to provide a cost-efficient answer. Experience, communication skills, and awareness of the special needs of the nonprofit institution are aspects of the consultant's advantages that the institutional manager needs to evaluate.

16.13   COST CONSIDERATIONS

Whenever there is a perceived need to use a contractor, it is always a good idea to perform an assessment to determine if one is *really* needed. Sometimes you will find that a consultant will cost more than developing the expertise in-house. However, this is usually not the case.

I. *Needs Assessment:* Do we need a consultant?
   A. What laws are applicable?
   B. What laws affect employees?
   C. What laws affect processes?
II. *Scope of Operations*
   A. small operation
   B. large operation
   C. single location
   D. multiple locations
   E. remote locations
   F. length of time to compliance
   G. one-time
   H. ongoing
III. *Can in-house expertise provide adequate coverage?*
   A. full-time or part-time individual(s)
   B. in-house
   C. outside consultants
   D. combination
IV. *Basic costs of maintaining in-house occupational-safety-and-health individual*
   A. five years experience: $35,000 to $60,000

      B. plus benefits: $49,000 to $70,000
      C. each entry-level tech: $25,000
      D. plus benefits: $35,000
   V. *Maintaining expertise in rapidly changing field*
      A. current literature on occupational safety and health: $2,000/year
      B. seminars/workshops: $2,000/year/person
      C. professional memberships and affiliations: $500/person/year
  VI. *How do you get management to pay for/support your programs?*
      A. compliance with laws
      B. liability
      C. save money by developing programs to decrease insurance costs

When using consultants it is very important to be specific as to your needs. Everything should be clear up front on what the service will be and where the end point will be. Many times, in the hopes of getting more business, if a specific scope of work is not described, a consultant will come up with broad recommendations that are not very realistic.

If the consultant's final report should be kept confidential, for example, as an environmental-compliance audit, then preparations for confidentiality should be worked out with your attorneys before the consultant is engaged. One way to do this is to have the report requested and paid for by the institution's attorney so that a future litigation request for disclosure may be avoided by the claim of "attorney work product privilege." In the case of environmental audits, state laws in a dozen or more states provide an explicit protection for the contents and recommendations of such an audit.

16.14   SAFETY MANAGEMENT'S RELATIONSHIPS WITH LEGAL COUNSEL

If your institution does not have in-house counsel, it is a prudent idea to develop a relationship with an attorney or law firm who is knowledgeable in the safety and environmental field and who knows your locality's regulators. Safety and environmental managers will often know the laws and regulations but can always use help with interpretation and application of the rules to specific situations.

In the environmental field, if you need to look for an attorney because of a legal problem, you are probably already in over your head. If your institution has in-house counsel, it would be a good idea to encourage their attendance at some safety and environmental seminars. It is always helpful to have the in-house counsel supporting your requests for more money and a larger staff.

Index